genetics

genetics

VOLUME 4
Q–Z

Richard Robinson

MACMILLAN
REFERENCE
USA™

THOMSON
✳
™
GALE

New York • Detroit • San Diego • San Francisco • Cleveland • New Haven, Conn. • Waterville, Maine • London • Munich

Genetics
Richard Robinson

Volume ISBN Numbers
0-02-865607-5 (Volume 1)
0-02-865608-3 (Volume 2)
0-02-865609-1 (Volume 3)
0-02-865610-5 (Volume 4)

LIBRARY OF CONGRESS CATALOGING- IN-PUBLICATION DATA

Genetics / Richard Robinson, editor in chief.
    p. ; cm.
Includes bibliographical references and index.
    ISBN 0-02-865606-7 (set : hd.)
    1. Genetics—Encyclopedias.
    [DNLM: 1. Genetics—Encyclopedias—English. 2. Genetic Diseases, Inborn—Encyclopedias—English. 3. Genetic Techniques—Encyclopedias—English. 4. Molecular Biology—Encyclopedias—English. QH 427 G328 2003]   I. Robinson, Richard, 1956–
    QH427 .G46 2003
    576'.03—dc21

                     2002003560

Printed in Canada
10 9 8 7 6 5 4 3 2 1

For Your Reference

The following section provides a group of diagrams and illustrations applicable to many entries in this encyclopedia. The molecular structures of DNA and RNA are provided in detail in several different formats, to help the student understand the structures and visualize how these molecules combine and interact. The full set of human chromosomes are presented diagrammatically, each of which is shown with a representative few of the hundreds or thousands of genes it carries.

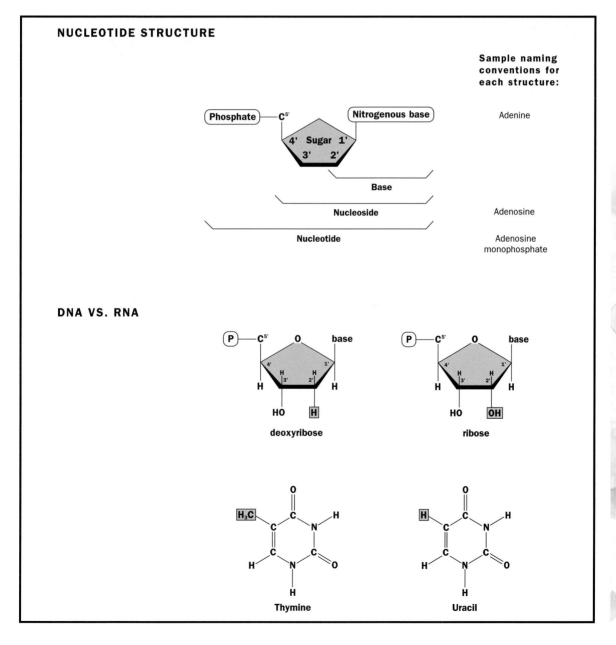

NUCLEOTIDE STRUCTURE

Sample naming conventions for each structure:

Phosphate — C5'

Nitrogenous base

4' Sugar 1'
3' 2'

Base

Nucleoside

Nucleotide

Adenine

Adenosine

Adenosine monophosphate

DNA VS. RNA

deoxyribose

ribose

Thymine

Uracil

NUCLEOTIDE STRUCTURES

Purine-containing DNA nucleotides

Adenine

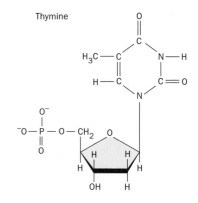

Guanine

Pyrimidine-containing DNA nucleotides

Thymine

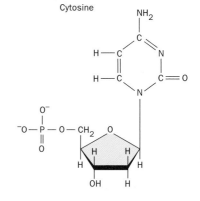

Cytosine

CANONICAL B-DNA DOUBLE HELIX

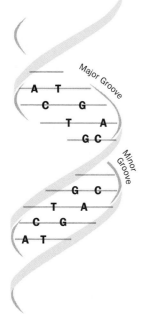

Ribbon model

Ball-and-stick model

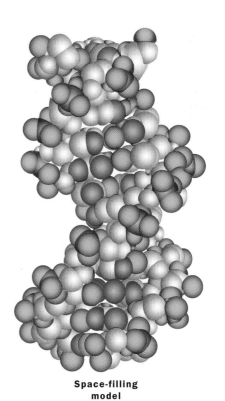

Space-filling model

DNA NUCLEOTIDES PAIR UP ACROSS THE DOUBLE HELIX; THE TWO STRANDS RUN ANTI-PARALLEL

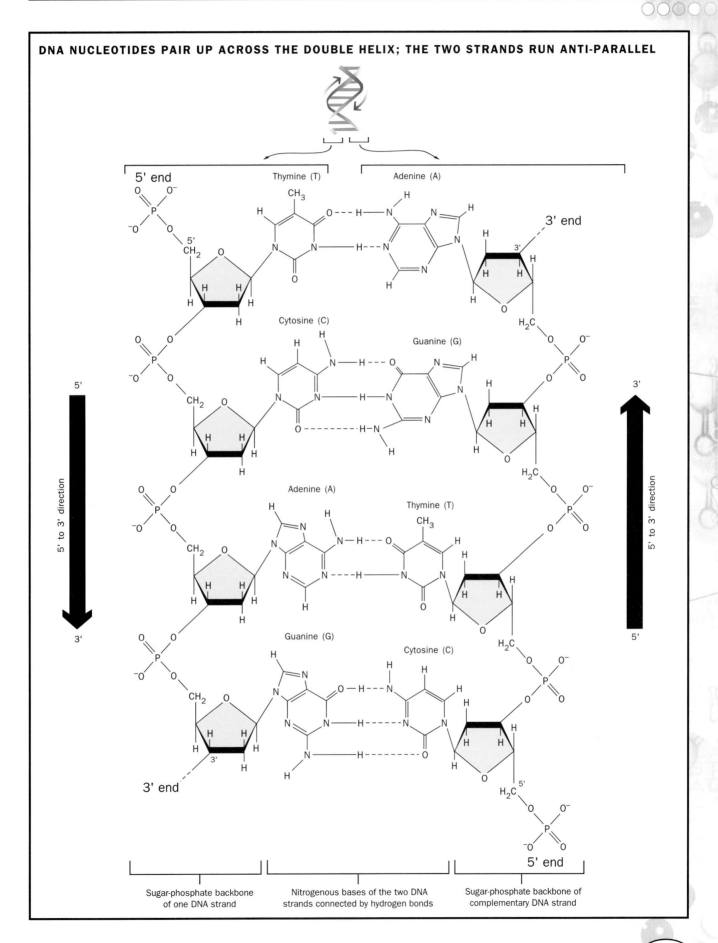

Sugar-phosphate backbone of one DNA strand

Nitrogenous bases of the two DNA strands connected by hydrogen bonds

Sugar-phosphate backbone of complementary DNA strand

SELECTED LANDMARKS OF THE HUMAN GENOME

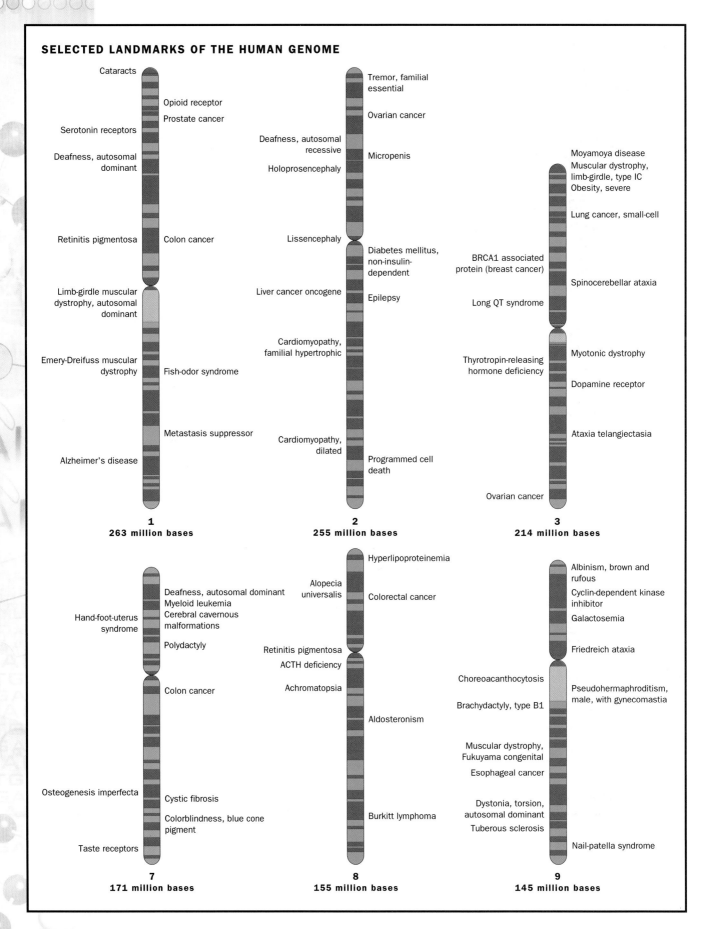

Cataracts

Opioid receptor
Prostate cancer

Serotonin receptors

Deafness, autosomal dominant

Retinitis pigmentosa — Colon cancer

Limb-girdle muscular dystrophy, autosomal dominant

Emery-Dreifuss muscular dystrophy — Fish-odor syndrome

Metastasis suppressor

Alzheimer's disease

1
263 million bases

Tremor, familial essential

Ovarian cancer

Deafness, autosomal recessive — Micropenis

Holoprosencephaly

Lissencephaly — Diabetes mellitus, non-insulin-dependent

Liver cancer oncogene — Epilepsy

Cardiomyopathy, familial hypertrophic

Cardiomyopathy, dilated — Programmed cell death

2
255 million bases

Moyamoya disease
Muscular dystrophy, limb-girdle, type IC
Obesity, severe

Lung cancer, small-cell

BRCA1 associated protein (breast cancer) — Spinocerebellar ataxia

Long QT syndrome

Thyrotropin-releasing hormone deficiency — Myotonic dystrophy

Dopamine receptor

Ataxia telangiectasia

Ovarian cancer

3
214 million bases

Hand-foot-uterus syndrome — Deafness, autosomal dominant
Myeloid leukemia
Cerebral cavernous malformations

Polydactyly

Colon cancer

Osteogenesis imperfecta — Cystic fibrosis

Colorblindness, blue cone pigment

Taste receptors

7
171 million bases

Hyperlipoproteinemia

Alopecia universalis — Colorectal cancer

Retinitis pigmentosa
ACTH deficiency
Achromatopsia

Aldosteronism

Burkitt lymphoma

8
155 million bases

Albinism, brown and rufous
Cyclin-dependent kinase inhibitor
Galactosemia

Friedreich ataxia

Choreoacanthocytosis — Pseudohermaphroditism, male, with gynecomastia

Brachydactyly, type B1

Muscular dystrophy, Fukuyama congenital
Esophageal cancer

Dystonia, torsion, autosomal dominant
Tuberous sclerosis

Nail-patella syndrome

9
145 million bases

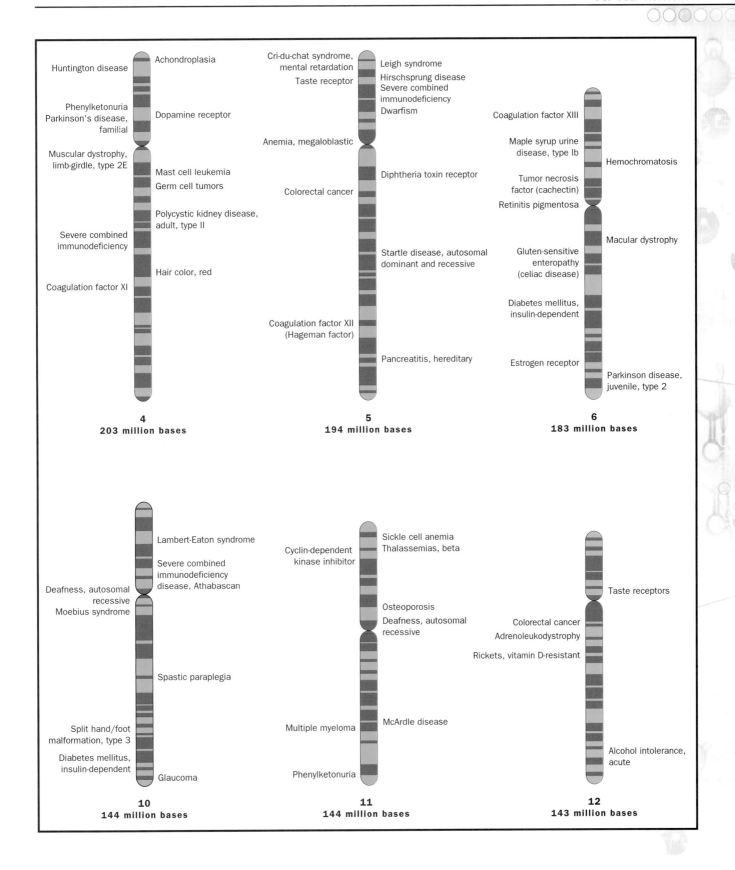

Huntington disease
Achondroplasia

Phenylketonuria
Parkinson's disease, familial
Dopamine receptor

Muscular dystrophy, limb-girdle, type 2E

Mast cell leukemia
Germ cell tumors

Polycystic kidney disease, adult, type II

Severe combined immunodeficiency

Hair color, red

Coagulation factor XI

4
203 million bases

Cri-du-chat syndrome, mental retardation
Taste receptor
Leigh syndrome
Hirschsprung disease
Severe combined immunodeficiency
Dwarfism

Anemia, megaloblastic

Diphtheria toxin receptor

Colorectal cancer

Startle disease, autosomal dominant and recessive

Coagulation factor XII (Hageman factor)

Pancreatitis, hereditary

5
194 million bases

Coagulation factor XIII

Maple syrup urine disease, type Ib
Hemochromatosis

Tumor necrosis factor (cachectin)
Retinitis pigmentosa

Macular dystrophy

Gluten-sensitive enteropathy (celiac disease)

Diabetes mellitus, insulin-dependent

Estrogen receptor
Parkinson disease, juvenile, type 2

6
183 million bases

Lambert-Eaton syndrome

Severe combined immunodeficiency disease, Athabascan

Deafness, autosomal recessive
Moebius syndrome

Spastic paraplegia

Split hand/foot malformation, type 3
Diabetes mellitus, insulin-dependent
Glaucoma

10
144 million bases

Cyclin-dependent kinase inhibitor
Sickle cell anemia
Thalassemias, beta

Osteoporosis
Deafness, autosomal recessive

Multiple myeloma
McArdle disease

Phenylketonuria

11
144 million bases

Taste receptors

Colorectal cancer
Adrenoleukodystrophy
Rickets, vitamin D-resistant

Alcohol intolerance, acute

12
143 million bases

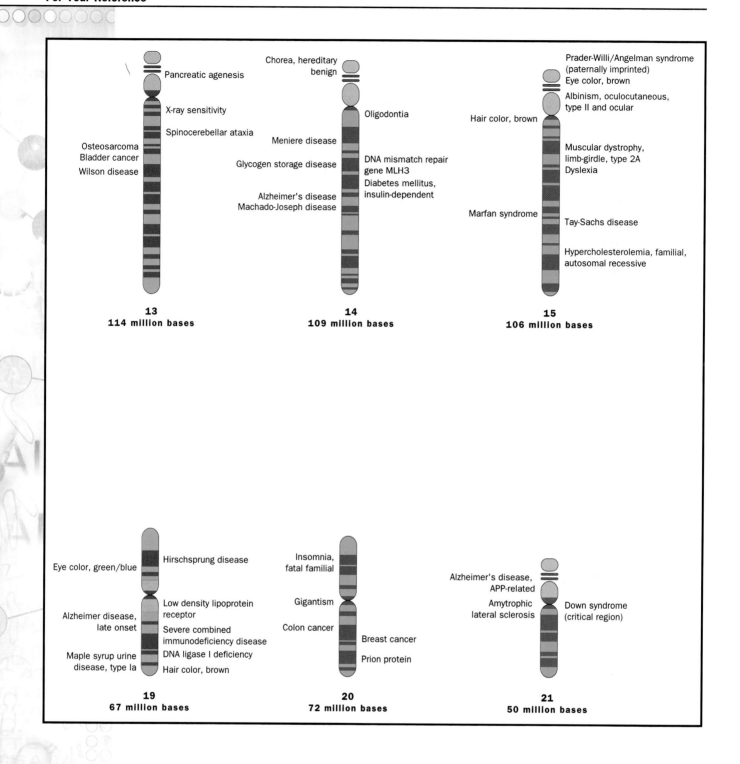

13
114 million bases

Pancreatic agenesis
X-ray sensitivity
Spinocerebellar ataxia
Osteosarcoma
Bladder cancer
Wilson disease

14
109 million bases

Chorea, hereditary benign
Oligodontia
Meniere disease
Glycogen storage disease
DNA mismatch repair gene MLH3
Diabetes mellitus, insulin-dependent
Alzheimer's disease
Machado-Joseph disease

15
106 million bases

Prader-Willi/Angelman syndrome (paternally imprinted)
Eye color, brown
Albinism, oculocutaneous, type II and ocular
Hair color, brown
Muscular dystrophy, limb-girdle, type 2A
Dyslexia
Marfan syndrome
Tay-Sachs disease
Hypercholesterolemia, familial, autosomal recessive

19
67 million bases

Eye color, green/blue
Hirschsprung disease
Low density lipoprotein receptor
Alzheimer disease, late onset
Severe combined immunodeficiency disease
Maple syrup urine disease, type Ia
DNA ligase I deficiency
Hair color, brown

20
72 million bases

Insomnia, fatal familial
Gigantism
Colon cancer
Breast cancer
Prion protein

21
50 million bases

Alzheimer's disease, APP-related
Amytrophic lateral sclerosis
Down syndrome (critical region)

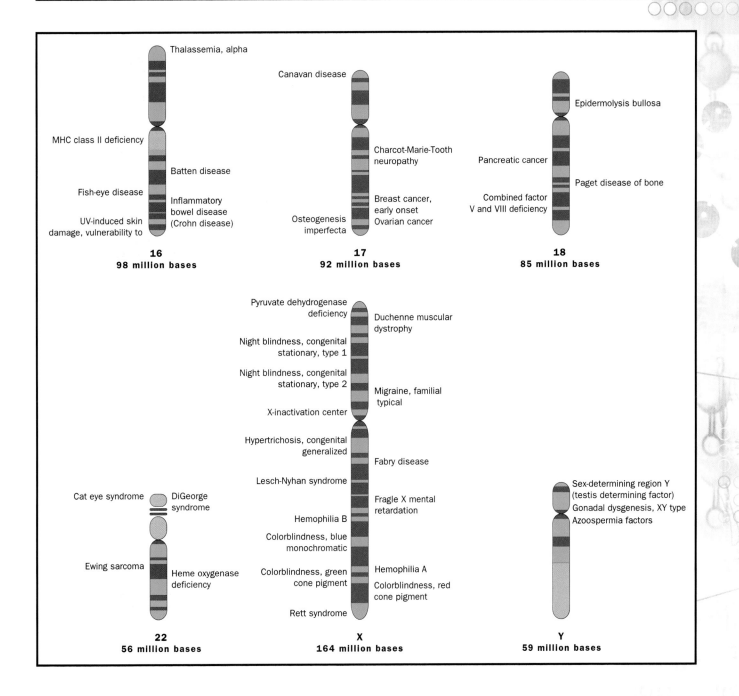

16
98 million bases

Thalassemia, alpha

MHC class II deficiency

Fish-eye disease

UV-induced skin damage, vulnerability to

Batten disease

Inflammatory bowel disease (Crohn disease)

17
92 million bases

Canavan disease

Charcot-Marie-Tooth neuropathy

Breast cancer, early onset

Ovarian cancer

Osteogenesis imperfecta

18
85 million bases

Epidermolysis bullosa

Pancreatic cancer

Paget disease of bone

Combined factor V and VIII deficiency

22
56 million bases

Cat eye syndrome

DiGeorge syndrome

Ewing sarcoma

Heme oxygenase deficiency

X
164 million bases

Pyruvate dehydrogenase deficiency

Night blindness, congenital stationary, type 1

Night blindness, congenital stationary, type 2

X-inactivation center

Hypertrichosis, congenital generalized

Lesch-Nyhan syndrome

Hemophilia B

Colorblindness, blue monochromatic

Colorblindness, green cone pigment

Rett syndrome

Duchenne muscular dystrophy

Migraine, familial typical

Fabry disease

Fragle X mental retardation

Hemophilia A

Colorblindness, red cone pigment

Y
59 million bases

Sex-determining region Y (testis determining factor)

Gonadal dysgenesis, XY type

Azoospermia factors

Contributors

Eric Aamodt
Louisiana State University Health Sciences Center, Shreveport
Gene Expression: Overview of Control

Maria Cristina Abilock
Applied Biosystems
Automated Sequencer
Cycle Sequencing
Protein Sequencing
Sequencing DNA

Ruth Abramson
University of South Carolina School of Medicine
Intelligence
Psychiatric Disorders
Sexual Orientation

Stanley Ambrose
University of Illinois
Population Bottleneck

Allison Ashley-Koch
Duke Center for Human Genetics
Disease, Genetics of
Fragile X Syndrome
Geneticist

David T. Auble
University of Virginia Health System
Transcription

Bruce Barshop
University of California, San Diego
Metabolic Disease

Mark A. Batzer
Louisiana State University
Pseudogenes
Repetitive DNA Elements
Transposable Genetic Elements

Robert C. Baumiller
Xavier University
Reproductive Technology
Reproductive Technology: Ethical Issues

Mary Beckman
Idaho Falls, Idaho
DNA Profiling
HIV

Samuel E. Bennett
Oregon State University Department of Genetics
DNA Repair
Laboratory Technician
Molecular Biologist

Andrea Bernasconi
Cambridge University, U.K.
Multiple Alleles
Nondisjunction

C. William Birky, Jr.
University of Arizona
Inheritance, Extranuclear

Joanna Bloom
New York University Medical Center
Cell Cycle

Deborah Blum
University of Wisconsin, Madison
Science Writer

Bruce Blumberg
University of California, Irvine
Hormonal Regulation

Suzanne Bradshaw
University of Cincinnati
Transgenic Animals
Yeast

Carolyn J. Brown
University of British Columbia
Mosaicism

Michael J. Bumbulis
Baldwin-Wallace College
Blotting

Michael Buratovich
Spring Arbor College
Operon

Elof Carlson
The State Universtiy of New York, Stony Brook
Chromosomal Theory of Inheritance, History
Gene
Muller, Hermann
Polyploidy
Selection

Regina Carney
Duke University
College Professor

Shu G. Chen
Case Western Reserve University
Prion

Gwen V. Childs
University of Arkansas for Medical Sciences
In situ Hybridization

Cindy T. Christen
Iowa State University
Technical Writer

Patricia L. Clark
University of Notre Dame
Chaperones

Steven S. Clark
University of Wisconsin
Oncogenes

Nathaniel Comfort
George Washington University
McClintock, Barbara

P. Michael Conneally
Indiana University School of Medicine
Blood Type
Epistasis
Heterozygote Advantage

Howard Cooke
Western General Hospital: MRC Human Genetics Unit
Chromosomes, Artificial

Denise E. Costich
Boyce Thompson Institute
Maize

Terri Creeden
March of Dimes
Birth Defects

Kenneth W. Culver
Novartis Pharmaceuticals Corporation
Genomics
Genomics Industry
Pharmaceutical Scientist

Mary B. Daly
Fox Chase Cancer Center
Breast Cancer

Pieter de Haseth
Case Western Reserve University
Transcription

Rob DeSalle
American Museum of Natural History
Conservation Geneticist
Conservation Biology: Genetic Approaches

Elizabeth A. De Stasio
Lawerence University
Cloning Organisms

Danielle M. Dick
Indiana University
Behavior

Michael Dietrich
Dartmouth College
Nature of the Gene, History

Christine M. Disteche
University of Washington
X Chromosome

Gregory Michael Dorr
University of Alabama
Eugenics

Jennie Dusheck
Santa Cruz, California
Population Genetics

Susanne D. Dyby
U.S. Department of Agriculture: Center for Medical, Agricultural, and Veterinary Entomology
Classical Hybrid Genetics
Mendelian Genetics
Pleiotropy

Barbara Emberson Soots
Folsom, California
Agricultural Biotechnology

Susan E. Estabrooks
Duke Center for Human Genetics
Fertilization
Genetic Counselor
Genetic Testing

Stephen V. Faraone
Harvard Medical School
Attention Deficit Hyperactivity Disorder

Gerald L. Feldman
Wayne State University Center for Molecular Medicine and Genetics
Down Syndrome

Linnea Fletcher
Bio-Link South Central Regional Coordinater, Austin Community College
Educator
Gel Electrophoresis
Marker Systems
Plasmid

Michael Fossel
Executive Director, American Aging Association
Accelerated Aging: Progeria

Carol L. Freund
National Institute of Health: Warren G. Magnuson Clinical Center
Genetic Testing: Ethical Issues

Joseph G. Gall
Carnegie Institution
Centromere

Darrell R. Galloway
The Ohio State University
DNA Vaccines

Pierluigi Gambetti
Case Western Reserve University
Prion

Robert F. Garry
Tulane University School of Medicine
Retrovirus
Virus

Perry Craig Gaskell, Jr.
Duke Center for Human Genetics
Alzheimer's Disease

Theresa Geiman
National Institute of Health: Laboratory of Receptor Biology and Gene Expression
Methylation

Seth G. N. Grant
University of Edinburgh
Embryonic Stem Cells
Gene Targeting
Rodent Models

Roy A. Gravel
University of Calgary
Tay-Sachs Disease

Nancy S. Green
March of Dimes
Birth Defects

Wayne W. Grody
UCLA School of Medicine
Cystic Fibrosis

Charles J. Grossman
Xavier University
Reproductive Technology
Reproductive Technology: Ethical Issues

Cynthia Guidi
University of Massachusetts Medical School
Chromosome, Eukaryotic

Patrick G. Guilfoile
Bemidji State University
DNA Footprinting
Microbiologist
Recombinant DNA
Restriction Enzymes

Richard Haas
University of California Medical Center
Mitochondrial Diseases

William J. Hagan
College of St. Rose
Evolution, Molecular

Jonathan L. Haines
Vanderbilt University Medical Center
Complex Traits
Human Disease Genes, Identification of

Mapping
McKusick, Victor

Michael A. Hauser
Duke Center for Human Genetics
DNA Microarrays
Gene Therapy

Leonard Hayflick
University of California
Telomere

Shaun Heaphy
University of Leicester, U.K.
Viroids and Virusoids

John Heddle
York University
Mutagenesis
Mutation
Mutation Rate

William Horton
Shriners Hospital for Children
Growth Disorders

Brian Hoyle
Square Rainbow Limited
Overlapping Genes

Anthony N. Imbalzano
University of Massachusetts Medical School
Chromosome, Eukaryotic

Nandita Jha
University of California, Los Angeles
Triplet Repeat Disease

John R. Jungck
Beloit College
Gene Families

Richard Karp
Department of Biological Sciences, University of Cincinnati
Transplantation

David H. Kass
Eastern Michigan University
Pseudogenes
Transposable Genetic Elements

Michael L. Kochman
University of Pennsylvania Cancer Center
Colon Cancer

Bill Kraus
Duke University Medical Center
Cardiovascular Disease

Steven Krawiec
Lehigh University
Genome

Mark A. Labow
Novartis Pharmaceuticals Corporation
Genomics
Genomics Industry
Pharmaceutical Scientist

Ricki Lewis
McGraw-Hill Higher Education; The Scientist
Bioremediation
Biotechnology: Ethical Issues
Cloning: Ethical Issues

Genetically Modified Foods
Plant Genetic Engineer
Prenatal Diagnosis
Transgenic Organisms: Ethical
  Issues

Lasse Lindahl
University of Maryland, Baltimore
Ribozyme
RNA

David E. Loren
*University of Pennsylvania School of
Medicine*
Colon Cancer

Dennis N. Luck
Oberlin College
Biotechnology

Jeanne M. Lusher
*Wayne State University School of
Medicine; Children's Hospital of
Michigan*
Hemophilia

Kamrin T. MacKnight
*Medlen, Carroll, LLP: Patent,
Trademark and Copyright Attorneys*
Attorney
Legal Issues
Patenting Genes
Privacy

Jarema Malicki
Harvard Medical School
Zebrafish

Eden R. Martin
Duke Center for Human Genetics
Founder Effect
Inbreeding

William Mattox
*University of Texas/Anderson
Cancer Center*
Sex Determination

Brent McCown
University of Wisconsin
Transgenic Plants

Elizabeth C. Melvin
Duke Center for Human Genetics
Gene Therapy: Ethical Issues
Pedigree

Ralph R. Meyer
University of Cincinnati
Biotechnology and Genetic Engi-
  neering, History of
Chromosome, Eukaryotic
Genetic Code
Human Genome Project

Kenneth V. Mills
College of the Holy Cross
Post-translational Control

Jason H. Moore
Vanderbilt University Medical School
Quantitative Traits
Statistical Geneticist
Statistics

Dale Mosbaugh
*Oregon State University: Center for
Gene Research and Biotechnology*

DNA Repair
Laboratory Technician
Molecular Biologist

Paul J. Muhlrad
University of Arizona
Alternative Splicing
Apoptosis
Arabidopsis thaliana
Cloning Genes
Combinatorial Chemistry
Fruit Fly: *Drosophila*
Internet
Model Organisms
Pharmacogenetics and Pharma-
  cogenomics
Polymerase Chain Reaction

Cynthia A. Needham
*Boston University School of
Medicine*
Archaea
Conjugation
Transgenic Microorganisms

R. John Nelson
University of Victoria
Balanced Polymorphism
Gene Flow
Genetic Drift
Polymorphisms
Speciation

Carol S. Newlon
*University of Medicine and
Dentistry of New Jersey*
Replication

Sophia A. Oliveria
*Duke University Center for Human
Genetics*
Gene Discovery

Richard A. Padgett
Lerner Research Institute
RNA Processing

Michele Pagano
*New York University Medical
Center*
Cell Cycle

Rebecca Pearlman
Johns Hopkins University
Probability

Fred W. Perrino
*Wake Forest University School of
Medicine*
DNA Polymerases
Nucleases
Nucleotide

David Pimentel
*Cornell University: College of
Agriculture and Life Sciences*
Biopesticides

Toni I. Pollin
*University of Maryland School of
Medicine*
Diabetes

Sandra G. Porter
Geospiza, Inc.
Homology

Eric A. Postel
Duke University Medical Center
Color Vision
Eye Color

Prema Rapuri
Creighton University
HPLC: High-Performance Liq-
  uid Chromatography

Anthony J. Recupero
Gene Logic
Bioinformatics
Biotechnology Entrepreneur
Proteomics

Diane C. Rein
BioComm Consultants
Clinical Geneticist
Nucleus
Roundworm: *Caenorhabditis ele-
  gans*
Severe Combined Immune Defi-
  ciency

Jacqueline Bebout Rimmler
Duke Center for Human Genetics
Chromosomal Aberrations

Keith Robertson
*Epigenetic Gene Regulation and
Cancer Institute*
Methylation

Richard Robinson
Tucson, Arizona
Androgen Insensitivity Syndrome
Antisense Nucleotides
Cell, Eukaryotic
Crick, Francis
Delbrück, Max
Development, Genetic Control of
DNA Structure and Function,
  History
Eubacteria
Evolution of Genes
Hardy-Weinberg Equilibrium
High-Throughput Screening
Immune System Genetics
Imprinting
Inheritance Patterns
Mass Spectrometry
Mendel, Gregor
Molecular Anthropology
Morgan, Thomas Hunt
Mutagen
Purification of DNA
RNA Interference
RNA Polymerases
Transcription Factors
Twins
Watson, James

Richard J. Rose
Indiana University
Behavior

Howard C. Rosenbaum
*Science Resource Center, Wildlife
Conservation Society*
Conservation Geneticist
Conservation Biology: Genetic
  Approaches

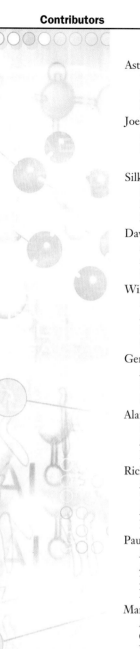

Astrid M. Roy-Engel
Tulane University Health Sciences Center
Repetitive DNA Elements

Joellen M. Schildkraut
Duke University Medical Center
Public Health, Genetic Techniques in

Silke Schmidt
Duke Center for Human Genetics
Meiosis
Mitosis

David A. Scicchitano
New York University
Ames Test
Carcinogens

William K. Scott
Duke Center for Human Genetics
Aging and Life Span
Epidemiologist
Gene and Environment

Gerry Shaw
MacKnight Brain Institute of the University of Flordia
Signal Transduction

Alan R. Shuldiner
University of Maryland School of Medicine
Diabetes

Richard R. Sinden
Institute for Biosciences and Technology: Center for Genome Research
DNA

Paul K. Small
Eureka College
Antibiotic Resistance
Proteins
Reading Frame

Marcy C. Speer
Duke Center for Human Genetics
Crossing Over
Founder Effect
Inbreeding
Individual Genetic Variation
Linkage and Recombination

Jeffrey M. Stajich
Duke Center for Human Genetics
Muscular Dystrophy

Judith E. Stenger
Duke Center for Human Genetics
Computational Biologist
Information Systems Manager

Frank H. Stephenson
Applied Biosystems
Automated Sequencer
Cycle Sequencing
Protein Sequencing
Sequencing DNA

Gregory Stewart
State University of West Georgia
Transduction
Transformation

Douglas J. C. Strathdee
University of Edinburgh
Embryonic Stem Cells
Gene Targeting
Rodent Models

Jeremy Sugarman
Duke University Department of Medicine
Genetic Testing: Ethical Issues

Caroline M. Tanner
Parkinson's Institute
Twins

Alice Telesnitsky
University of Michigan
Reverse Transcriptase

Daniel J. Tomso
National Institute of Environmental Health Sciences
DNA Libraries
Escherichia coli
Genetics

Angela Trepanier
Wayne State University Genetic Counseling Graduate Program
Down Syndrome

Peter A. Underhill
Stanford University
Y Chromosome

Joelle van der Walt
Duke University Center for Human Genetics
Genotype and Phenotype

Jeffery M. Vance
Duke University Center for Human Genetics

Gene Discovery
Genomic Medicine
Genotype and Phenotype
Sanger, Fred

Gail Vance
Indiana University
Chromosomal Banding

Jeffrey T. Villinski
University of Texas/MD Anderson Cancer Center
Sex Determination

Sue Wallace
Santa Rosa, California
Hemoglobinopathies

Giles Watts
Children's Hospital Boston
Cancer
Tumor Suppressor Genes

Kirk Wilhelmsen
Ernest Gallo Clinic & Research Center
Addiction

Michelle P. Winn
Duke University Medical Center
Physician Scientist

Chantelle Wolpert
Duke University Center for Human Genetics
Genetic Counseling
Genetic Discrimination
Nomenclature
Population Screening

Harry H. Wright
University of South Carolina School of Medicine
Intelligence
Psychiatric Disorders
Sexual Orientation

Janice Zengel
University of Maryland, Baltimore
Ribosome
Translation

Stephan Zweifel
Carleton College
Mitochondrial Genome

Table of Contents

genetics

Quantitative Traits

Quantitative traits are those that vary continuously. This is in contrast to qualitative traits, in which the **phenotype** is discrete and can take on one of only a few different values. Examples of quantitative traits include height, weight, and blood pressure. There is no single gene for any of these traits, instead it is generally believed that continuous variation in a trait such as blood pressure is partly due to DNA sequence variations at multiple genes, or **loci**. Such loci are referred to as quantitative trait loci (QTL). Much of how we study and characterize quantitative traits can be attributed to the work of Ronald Fisher and Sewall Wright, accomplished during the first half of the twentieth century.

phenotype observable characteristics of an organism

loci site on a chromosome (singular, locus)

The Genetic Architecture of Quantitative Traits

An important goal of genetic studies is to characterize the genetic architecture of quantitative traits. Genetic architecture can been defined in one of four ways. First, it refers to the number of QTLs that influence a quantitative trait. Second, it can mean the number of **alleles** that each QTL has. Third, it reflects the frequencies of the alleles in the population. And fourth, it refers to the influence of each QTL and its alleles on the quantitative trait. Imagine, for instance, a quantitative trait influenced by 6 loci, each of which has 3 alleles. This gives a total of 18 possible allele combinations. Some alleles may be very rare in a population, so that the phenotypes it contributes to may be rare as well. Some alleles have disproportionate effects on the phenotype (for instance, an allele that causes dwarfism), which may mask the more subtle effects of other alleles. The trait may also be influenced by the environment, giving an even wider range of phenotypic possibilities.

alleles particular forms of genes

Understanding the genetic architecture of quantitative traits is important in a number of disciplines, including animal and plant breeding, medicine, and evolution. For example, a quantitative trait of interest to animal breeders might be meat quality in pigs. The identification and characterization of QTLs for meat quality might provide a basis for selecting and breeding pigs with certain desirable features. In medicine, an important goal is to identify genetic risk factors for various common diseases. Many genetic studies of common disease focus on the presence or absence of disease as the trait of interest. In some cases, however, quantitative traits may provide more information for identifying genes than qualitative traits. For example,

Human height is a quantitative trait, controlled by multiple genes. The broad distribution of heights reflects this fact. Note that most people have an intermediate height, a typical distribution pattern for quantitative traits.

genotypes sets of genes present

identifying genetic risk factors for cardiovascular disease might be facilitated by studying the genetic architecture of cholesterol metabolism or blood pressure rather that the presence or absence of cardiovascular disease itself. Cholesterol metabolism is an example of an intermediate trait or endophenotype for cardiovascular disease. That is, it is related to the disease and may be useful as a "proxy measure" of the disease.

QTLs and Complex Effects on Phenotype

It is important to note that QTLs can influence quantitative traits in a number of different ways. First, variation at a QTL can impact quantitative trait levels. That is, the average or mean of the observed phenotypes for the trait may be different among different **genotypes** (for example, some genotypes will produce taller organisms than others). This is important because much of the basic theory underlying statistical methods for studying quantitative traits is based on genotypic means. For this reason, most genetic studies focus on quantitative trait means. However, there are a variety of other ways QTLs can influence quantitative traits. For example, it is possible that the trait means are the same among different genotypes but that the variances (the spread on either side of the mean) are not. In other words, variation in phenotypic values may be greater for some genotypes than for others—some genotypes, for example, may give a wider range of heights than others. This is believed to be due to gene-gene and gene-environment interactions such that the magnitude of the effects of a particular environmental or genetic factor may differ across genotypes.

It is also possible for QTLs to influence the relationship or correlation among quantitative traits. For example, the rate at which two proteins bind might be due to variation in the QTLs that code for those proteins. As a final example, QTLs can also impact the dynamics of a trait. That is, change in a phenotype over time might be due to variation at a

QTL, such as when blood pressure varies with the age of the individual. Thus, QTLs can affect quantitative trait levels, variability, co-variability, and dynamics.

In addition, each type of QTL effect may depend on a particular genetic or environmental context. Thus, the influence of a particular QTL on quantitative trait levels, variability, covariability, or dynamics may depend on one or more other QTLs (an effect called **epistasis** or gene-gene interaction) and/or one or more environmental factors. Although such context-dependent effects may be very common, and may play an important role in genetic architecture, they are typically very difficult to detect and characterize. This is partly due to limits of available statistical methods and the availability of large sample sizes.

Analysis of Quantitative Traits

Characterization of the genetic architecture of quantitative traits is typically carried out using one of two different study designs. The first approach starts with the quantitative trait of interest (such as height or blood pressure) and attempts to draw inferences about the underlying genetics from looking at the degree of trait resemblance among related subjects. This approach is sometimes referred to as a top-down or unmeasured genotype strategy because the inheritance pattern of the trait is the focus and no genetic variations are actually measured. The top-down approach is often the first step taken to determine whether there is evidence for a genetic component.

Heritability (the likelihood that the trait will be passed on to offspring) and **segregation analysis** are examples of statistical analyses that use a top-down approach. With the bottom-up or measured genotype approach, candidate QTLs are measured and then used to draw inferences about which genes might play a role in the genetic architecture of a quantitative trait. Prior to the availability of technologies for measuring QTLs, the top-down approach was very common. However, it is now inexpensive and efficient to measure many QTLs, making the bottom-up strategy a common study design. **Linkage analysis** and **association analysis** are two general statistical approaches that utilize the bottom-up study design.

The definition and characterization of quantitative traits is changing very rapidly. New technologies such as DNA microarrays and protein mass spectrometry are making it possible to quantitatively measure the expression levels of thousands of genes simultaneously. These new measures make it possible to study gene expression at both the RNA level and the protein level as a quantitative trait. These new quantitative traits open the door for understanding the hierarchy of the relationship between QTL variation and variation in quantitative traits at both the biochemical and physiological level. SEE ALSO COMPLEX TRAITS; DNA MICROARRAYS; GENE DISCOVERY; LINKAGE AND RECOMBINATION.

Jason H. Moore

epistasis supression of a characteristic of one gene by the action of another gene

segregation analysis statistical test to determine pattern of inheritance for a trait

linkage analysis examination of co-inheritance of disease and DNA markers, used to locate disease genes

association analysis estimation of the relationship between alleles or genotypes and disease

Bibliography

Griffiths, A. J., et al. *An Introduction to Genetic Analysis.* New York: W. H. Freeman, 2000.

Hartl, D. L., and A. G. Clark. *Principle of Population Genetics.* Sunderland, MA: Sinauer Associates, 1997.

codon a sequence of three mRNA nucleotides coding for one amino acid

transcription messenger RNA formation from a DNA sequence

glycolipid molecule composed of sugar and fatty acid

Reading Frame

Almost all organisms translate their genes into protein structures using an identical, universal **codon** dictionary in which each amino acid in the protein is represented by a combination of only three nucleotides. For example, the sequence AAA in a gene is transcribed into the sequence UUU in messenger RNA (mRNA) and is then translated as the amino acid phenylalanine. A group of several codons that, taken together, provide the code for an amino acid, is called a reading frame. There are no "spaces" in the mRNA to denote the end of one codon and the start of another. Instead, the reading frame, or group of triplets, is determined solely by initial position of the pattern-making machinery at the start of the translation. In order for correct translation to occur, this reading frame must be maintained throughout the **transcription** and translation process.

Any single or double base insertions or deletions in the DNA or RNA sequence will throw off the reading frame and result in aberrant gene expression. Mutations that result in such insertions or deletions are termed "frameshift mutations." The insertion of three nucleotides, on the other hand, will only extend the length of the protein without affecting the reading frame, although it may affect the function of the protein. Several genetic diseases, including Huntington's disease, contain such trinucleotide repeats.

Because DNA consists of four possible bases and each codon consists of only a three-base sequence, there are 4^3, or sixty-four possible codons for the twenty common amino acids. In the codon dictionary, sixty-one of the codons code for amino acids, with the remaining three codons marking the end of the reading frame. The codon AUG denotes both the amino acid methionine and the start of the reading frame. In several cases, more than one codon can result in the creation of the same amino acid. For example CAC and CAU both code for histidine. This condition is termed "degeneracy," and it means that some mutations may still result in the same amino acid being inserted at that point into the protein. The above example also explains the "wobble hypothesis," put forward by Francis Crick, which states that substitutions in the terminal nucleotide of a codon have little or no effect on the proper insertion of amino acids during translation.

Medically important frameshift mutations include an insertion in the gene for a rare form of Gaucher disease preventing **glycolipid** breakdown. Charcot-Marie-Tooth disease, which results in numbness in hands and feet, is caused by the repetitive insertion of 1.5 million base pairs into the gene. A frameshift mutation of four bases in the gene coding for the low-density lipid receptor near one end causes the receptor to improperly anchor itself in the cell membrane, resulting in the faulty turnover of cholesterol that

Insertion of two Cs shifts the reading frame, creating a premature stop codon.

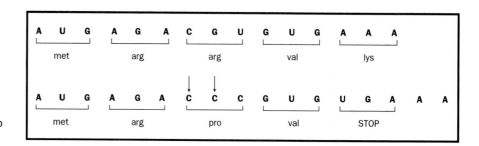

causes hypercholesteroiemia, or high blood levels of cholesterol. A single nucleotide pair deletion in codon 55 of the gene coding for phenylalanine hydroxylase (PAH) results in a form of phenylketonuria. Frameshift mutations are denoted by listing the location and specific change in the DNA. For example, 55delT indicates a thymidine was deleted in the 55th codon of the PAH gene. SEE ALSO CRICK, FRANCIS; GENETIC CODE; MUTATION; TRANSCRIPTION; TRANSLATION.

Paul K. Small

Three different reading frames for one mRNA sequence

Bibliography

Fairbanks, Daniel J., and W. Ralph Anderson. *Genetics: The Continuity of Life.* Pacific Grove, CA: Brooks/Cole, 1999.

Lewis, Ricki. *Human Genetics: Concepts and Applications,* 4th ed. New York: McGraw-Hill, 2001.

Lodish, Harvey, et al. *Molecular Cell Biology,* 4th ed. New York: W. H. Freeman, 2000.

Pasternak, Jack J. *Human Molecular Genetics: Mechanisms of Inherited Diseases.* Bethesda, MD: Fitzgerald Science Press, 1999.

Recessiveness   *See Inheritance Patterns*

Recombinant DNA

Recombinant DNA refers to a collection of techniques for creating (and analyzing) DNA molecules that contain DNA from two unrelated organisms. One of the DNA molecules is typically a bacterial or viral DNA that is capable of accepting another DNA molecule; this is called a **vector** DNA. The other DNA molecule is from an organism of interest, which could be anything from a bacterium to a whale, or a human. Combining these two DNA molecules allows for the **replication** of many copies of a specific DNA. These copies of DNA can be studied in detail, used to produce valuable proteins, or used for gene therapy or other applications.

vector carrier

replication duplication of DNA

The development of recombinant DNA tools and techniques in the early 1970s led to much concern about developing genetically modified organisms with unanticipated and potentially dangerous properties. This concern led to a proposal for a voluntary moratorium on recombinant DNA research in 1974, and to a meeting in 1975 at the Asilomar Conference Center in California. Participants at the Asilomar Conference agreed to a set of safety standards for recombinant DNA work, including the use of disabled bacteria that were unable to survive outside the laboratory. This conference helped satisfy the public about the safety of **recombinant DNA** research, and led to a rapid expansion of the use of these powerful new technologies.

recombinant DNA DNA formed by combining segments of DNA, usually from different types of organisms

Overview of Recombination Techniques

The basic technique of recombinant DNA involves digesting a vector DNA with a **restriction enzyme**, which is a molecular scissors that cuts DNA at specific sites. A DNA molecule from the organism of interest is also digested, in a separate tube, with the same restriction enzyme. The two DNAs are then mixed together and joined, this time using an enzyme called DNA ligase, to make an intact, double-stranded DNA molecule. This construct is

restriction enzyme an enzyme that cuts DNA at a particular sequence

then put into *Escherichia coli* cells, where the resulting DNA is copied billions of times. This novel DNA molecule is then isolated from the *E. coli* cells and analyzed to make sure that the correct construct was produced. This DNA can then be sequenced, used to generate protein from *E. coli* or another host, or for many other purposes.

There are many variations on this basic method of producing recombinant DNA molecules. For example, sometimes researchers are interested in isolating a whole collection of DNAs from an organism. In this case, they digest the whole **genome** with restriction enzyme, join many DNA fragments into many different vector molecules, and then transform those molecules into *E. coli*. The different *E. coli* cells that contain different DNA molecules are then pooled, resulting in a "library" of *E. coli* cells that contain, collectively, all of the genes present in the original organism.

genome the total genetic material in a cell or organism

Another variation is to make a library of all expressed genes (genes that are used to make proteins) from an organism or tissue. In this case, RNA is isolated. The isolated RNA is converted to DNA using the enzyme called reverse transcriptase. The resulting DNA copy, commonly abbreviated as cDNA, is then joined to vector molecules and put into *E. coli*. This collection of recombinant cDNAs (a cDNA library) allows researchers to study the expressed genes in an organism, independent from nonexpressed DNA.

Applications

Recombinant DNA technology has been used for many purposes. The Human Genome Project has relied on recombinant DNA technology to generate libraries of genomic DNA molecules. Proteins for the treatment or diagnosis of disease have been produced using recombinant DNA techniques. In recent years, a number of crops have been modified using these methods as well.

As of 2001, over eighty products that are currently used for treatment of disease or for vaccination had been produced using recombinant DNA techniques. The first was human insulin, which was produced in 1978. Other protein therapies that have been produced using recombinant DNA technology include hepatitis B vaccine, human growth hormone, clotting factors for treating hemophilia, and many other drugs. At least 350 additional recombinant-based drugs are currently being tested for safety and efficacy. In addition, a number of diagnostic tests for diseases, including tests for hepatitis and AIDS, have been produced with recombinant DNA technology.

Gene therapy is another area of applied genetics that requires recombinant DNA techniques. In this case, the recombinant DNA molecules themselves are used for therapy. Gene therapy is being developed or attempted for a number of inherited human diseases.

Recombinant DNA technology has also been used to produce genetically modified foods. These include tomatoes that can be vine-ripened before shipping and rice with improved nutritional qualities. Genetically modified foods have generated controversy, and there is an ongoing debate in some communities about the benefits and risks of developing crops using recombinant DNA technology.

Since the mid-1970s, recombinant DNA techniques have been widely applied in research laboratories and in pharmaceutical and agricultural companies. It is likely that this relatively new area of genetics will continue to play an increasingly important part in biological research into the foreseeable future. SEE ALSO BIOTECHNOLOGY; CLONING GENES; CROSSING OVER; DNA LIBRARIES; *ESCHERICHIA COLI*; GENE THERAPY: ETHICAL ISSUES; GENETICALLY MODIFIED FOODS; HUMAN GENOME PROJECT; PLASMID; RESTRICTION ENZYMES; REVERSE TRANSCRIPTASE; TRANSPOSABLE GENETIC ELEMENTS.

Patrick G. Guilfoile

Bibliography

Cooper, Geoffrey. *The Cell: A Molecular Approach*. Washington, DC: ASM Press, 1997.

Glick, Bernard, and Jack Pasternak. *Molecular Biotechnology: Principles and Applications of Recombinant DNA*, 2nd ed. Washington, DC: ASM Press, 1998.

Kreuzer, Helen, and Adrianne Massey. *Recombinant DNA and Biotechnology*, 2nd ed. Washington, DC: ASM Press, 2000.

Lodish, Harvey, et al. *Molecular Cell Biology*, 4th ed. New York: W. H. Freeman, 2000.

Old, R. W., and S. B. Primrose. *Principles of Gene Manipulation*, 5th ed. London: Blackwell Scientific Publications, 1994.

Internet Resource

"Approved Biotechnology Drugs." Biotechnology Industry Organization. <http://www.bio.org/aboutbio/guide2.html>.

Repetitive DNA Elements

The human genome contains approximately three billion **base pairs** of DNA. Within this there are between 30,000 and 70,000 genes, which together add up to less than 5 percent of the entire **genome**. Most of the rest is made up of several types of noncoding repeated elements.

Most gene sequences are unique, found only once in the genome. In contrast, repetitive DNA elements are found in multiple copies, in some cases thousands of copies, as shown in Table 1. Unlike genes, most repetitive elements do not code for protein or RNA. Repetitive elements have been found in most other eukaryotic genomes that have been analyzed. What functions they serve, if any, are mainly unknown. Their presence and spread causes several inherited diseases, and they have been linked to major events in evolution.

Types of Repetitive Elements

Repetitive elements differ in their position in the genome, sequence, size, number of copies, and presence or absence of coding regions within them. The two major classes of repetitive elements are interspersed elements and tandem arrays.

Interspersed repeated elements are usually present as single copies and distributed widely throughout the genome. The interspersed repeats alone constitute about 45 percent of the genome. The best-characterized interspersed repeats are the transposable genetic elements, also called mobile elements or "jumping genes" (Figure 1).

base pairs two nucleotides (either DNA or RNA) linked by weak bonds

genome the total genetic material in a cell or organism

DISPERSED REPEATS (MOBILE ELEMENTS)

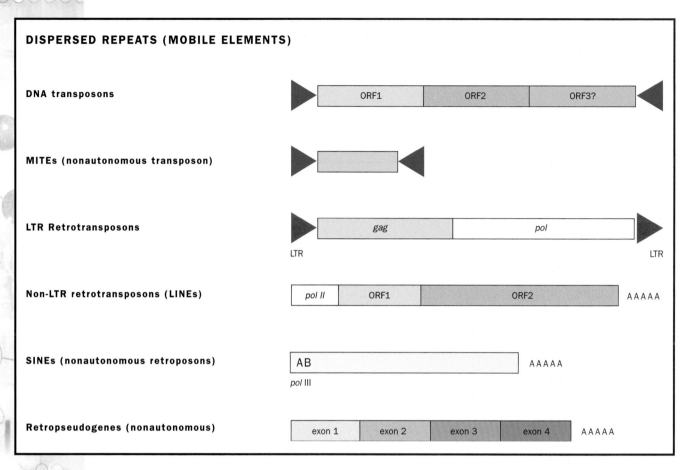

Figure 1. Dispersed repeats are usually found as single units spread throughout the genome. The DNA transposons have two or three open reading frames (ORFs) coding for the factors needed for amplification. LTR retrotransposons are flanked by long terminal repeats (LTR) and have genes such as *gag* (group antigen gene), and *pol* (polymerase gene). Non-LTR retrotransposons or LINEs will have ORFs and are transcribed by RNA polymerase II (*pol* II). There are two nonautonomous repeats thought to parasitize LINEs for its amplification: SINEs and retropseudogenes.

Sequences that are "tandemly arrayed" are present as duplicates, either head to tail or head to head. So-called satellites, minisatellites, and microsatellites largely exist in the form of tandem arrays (these elements originally got their name as "satellites" because they separated from the bulk of nuclear DNA during centrifugation). Sequences repeated in tandem are common at the centromere (where the two halves of a replicated chromosome are held together), and at or near the telomeres (the chromosome tips). Because they are difficult to sequence, sequences repeated in tandem at centromeres and telomeres are underrepresented in the draft sequence of the human genome. This makes it difficult to estimate the copy number, but they certainly represent at least 10 percent of the genome.

Tandem Arrays

Satellites (also called classical satellites), which occur in four classes (I–IV), form arrays of 1,000 to 10 million repeated units, particularly in the **heterochromatin** of chromosomes. They are concentrated in centromeres and account for much of the DNA there. Satellites of one type, called alphasatellites, occur as repeated units of approximately 171 base pairs (bp) in length, with high levels of sequence variation between the repeated units, as shown in Figure 2.

heterochromatin condensed portion of chromosomes

Table 1.

NUMBER OF COPIES OF INTERSPERSED REPEATS OBSERVED IN THE DRAFT OF THE HUMAN GENOME

Repeat type	Copy number	Fraction of the genome (%)
DNA transposons	294	3
LTR retrotransposons	443	8
Non-LTR retrotransposons:	2,426	34
LINEs	868	21
SINEs	1,558	13
Others	3	0.1

Minisatellites form arrays of several hundred units of 7 to 100 bp in length. They are present everywhere with an increasing concentration toward the telomeres. They differ from satellites in that they are found only in moderate numbers of tandem repeats and because of their high degree of dispersion throughout chromosomes.

Microsatellites, or simple sequence repeats (SSRs), are composed of units of one to six **nucleotides**, repeated up to a length of 100 bp or more. One-third are simple "polyadenylated" repeats, composed of nothing but adenine nucleotides. Other examples of abundant microsatellites are $(AC)_n$, $(AAAN)_n$, $(AAAAN)_n$, and $(AAN)_n$, where N represents any nucleotide and n is the number of repeats. Less abundant, but important because of their direct involvement in the generation of disease, are the $(CAG/CTG)_n$ and $(CGG/CCG)_n$ trinucleotide (or triplet) repeats.

nucleotides the building blocks of RNA or DNA

Telomeric and subtelomeric repeats are present at the end of the telomeres and are composed of short tandem repeats (STRs) of $(TTAGGG)_n$, up to 30,000 bp long. This sequence is "highly conserved," meaning it has changed very little over evolutionary time, indicating it likely plays a very important role. These STRs function as caps or ends of the long linear chromosomal DNA molecule and are crucial to the maintenance of intact eukaryotic chromosomes. Subtelomeric repeats act as transitions between the boundary of the telomere and the rest of the chromosome. They contain units similar to the TTAGGG, but they are not conserved.

Transposable Elements

Transposable elements are classified as either transposons or retrotransposons, depending on their mechanism of amplification. Transposons directly synthesize a DNA copy of themselves, whereas retrotransposons generate an RNA intermediate that is then reverse-transcribed (by the enzyme reverse transcriptase) back into DNA. Transposable elements fall into three major groups: DNA transposons, long terminal repeat (LTR) retrotransposons, and non-LTR retrotransposons. They also are subdivided into autonomous and nonautonomous elements, based on whether they can move independently within the genome or require other elements to perform this process, as shown in Figure 3.

DNA transposons are flanked by inverted repeats and contain two or more open reading frames (ORFs). An ORF is a DNA sequence that can be transcribed to make protein. The ORFs in DNA transposons code for the proteins required for making transposon copies and spreading them through the genome. The nonautonomous elements miniature inverted-repeat transposable elements (MITEs) are derived from a parent DNA transposon that

Figure 2. Schematic organization of the structures characteristic of one type of repeated sequences found in eukaryotic genomes. Tandem repeats consist of a different variety of sequence units (shown in parentheses) that are found repeated side by side, with the number of copies being very high.

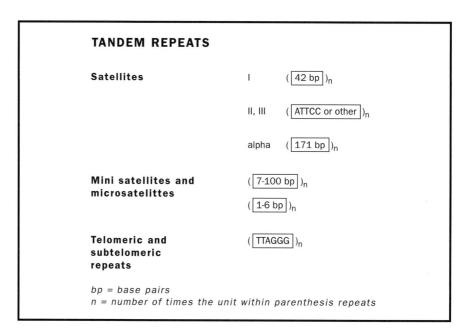

endoplasmic reticulum network of membranes within the cell

mRNA messenger RNA

promoter DNA sequence to which RNA polymerase binds to begin transcription

introns untranslated portions of genes that interrupt coding regions

lost ORF sequences, making them unable to amplify on their own. Instead, they must borrow the factors for amplification from external sources.

LTR retrotransposons are very similar to the genomes of retroviruses. They are flanked by 250 to 600 bp direct repeats called long terminal repeats. In general, not only are these elements defective, but they also appear to have deletions typical of nonautonomous families.

Several different groups of non-LTR retrotransposons can be found throughout most, if not all, eukaryotic genomes. One of these groups, the long interspersed repeated elements (LINEs), constitute about 21 percent of the human genome, with L1 and L2 being the dominant elements. Most of the element copies are incomplete and inactive. Two types of non-autonomous elements are thought to use factors made by LINEs: short interspersed repeated elements (SINEs) and retropseudogenes.

SINEs are derived from two types of genes coding for RNA: 7SL (which aids the movement of new proteins into the **endoplasmic reticulum**) and transfer RNAs. The most abundant human SINE is Alu, constituting about 13 percent of the human genome.

Retropseudogenes are derived from retrotransposition of **mRNA** derived from different genes. They can be distinguished from the parental gene by their lack of a functional **promoter** and by their lack of **introns**. The human genome is estimated to contain 35,000 copies of different retropseudogenes.

Role of Repetitive DNA in Evolution and Impact on the Human Genome

Most eukaryotic genomes contain repetitive DNA. Although most repeated sequences have no known function, their impact and importance on genomes is evident. Mobile repeated elements have been a critical factor in gene evolution. It has been suggested that some types of repeats may be linked to speciation, since during the evolutionary period when there was a high activity of mobile elements, radiation of different species occurred.

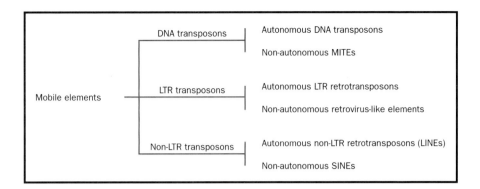

Figure 3. Representation of the division of the major types of mobile elements. Mobile elements can be subdivided into groups depending on their mechanism of amplification (DNA or RNA based), their structure (presence of LTR elements) and their ablilty for autonomous and nonautonomous amplification.

There are several diseases linked to—or caused by—repetitive elements. Expansion of triplet repeats has been tied to fragile X syndrome (a common cause of mental retardation), Huntington's disease, myotonic muscular dystrophy, and several other diseases. In addition, the discovery of STR instability in certain cancers suggests that sequence instability may play a role in cancer progression.

Mobile elements have caused diseases when a new mobile element disrupts an important gene. Neurofibromatosis type 1, for example, is caused by the insertion of an Alu element in the gene *NF1*. Alternatively, recombination between two repeated elements within a gene will alter its function, also causing disease. Many examples of cancers (e.g., acute myelogenous leukemia) and inherited diseases (e.g., alpha thalassemia) are caused by mobile-element-based recombinations.

Application of Repeats to Human Genomic Studies

Repeated sequences can be useful genetic tools. Because many of the repeated sequences are stably inherited, highly conserved, and found throughout the genome, they are ideal for genetic studies: They can act as "signposts" for finding and mapping functional genes. In addition, a repeat at a particular locus may be absent in one individual, or it may differ between two individuals (**polymorphism**). This makes repeats useful for identifying specific individuals (called DNA profiling) and their ancestors (molecular anthropology).

polymorphism DNA sequence variant

Microsatellites, in particular, have been used to identify individuals, study populations, and construct evolutionary trees. They have also been used as markers for disease-gene mapping and to evaluate specific genes in tumors. LINEs, and particularly the human SINE Alu, have been used for studies of human population genetics, primate comparative genomics, and DNA profiling. SEE ALSO CENTROMERE; CHROMOSOME, EUKARYOTIC; *IN SITU* HYBRIDIZATION; POLYMORPHISMS; PSEUDOGENES; RETROVIRUS; TELOMERE; TRANSPOSABLE GENETIC ELEMENTS; TRIPLET REPEAT DISEASE.

Astrid M. Roy-Engel and Mark A. Batzer

Bibliography

Deininger, Prescott L., and Mark A. Batzer. "Alu Repeats and Human Disease." *Molecular Genetics and Metabolism* 67, no. 3 (1999): 183–193.

Deininger, Prescott L., and Astrid M. Roy-Engel. "Mobile Elements in Animal and Plant Genomes." In *Mobile DNA II*, Nancy L. Craig, et al., eds. Washington, DC: ASM Press, 2001.

Figure 2. Model of a bacterial replication fork.

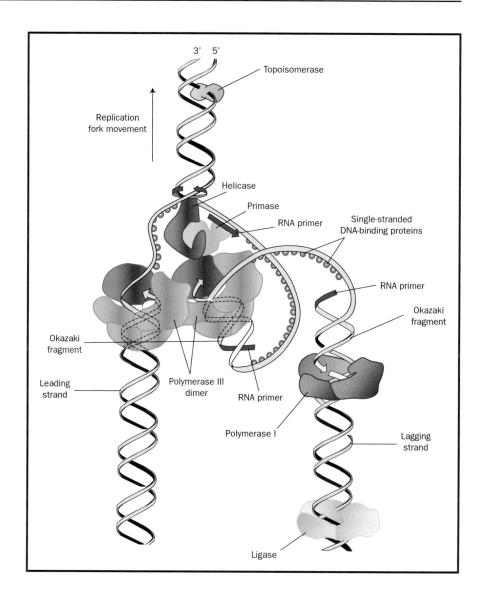

double-stranded DNA is continually unwound and copied. The unwinding of DNA poses special problems, which can be visualized by imagining pulling apart two pieces of string that are tightly wound around each other. The pulling apart requires energy; the strands tend to rewind if not held apart; and the region ahead of the separated strands becomes even more tightly twisted.

Proteins at the replication fork address each of these problems. DNA polymerases are not able to unwind double-stranded DNA, which requires energy to break the hydrogen bonds between the bases that hold the strands together. This task is accomplished by the enzyme DNA helicase, which uses the energy in **ATP** to unwind the template DNA at the replication fork. The single strands are then bound by a single-strand binding protein (called SSB in bacteria and RPA in eukaryotes), which prevents the strands from reassociating to form double-stranded DNA. Unwinding the DNA at the replication fork causes the DNA ahead of the fork to rotate and become twisted on itself. To prevent this from happening, an enzyme called DNA gyrase (in bacteria) or topoisomerase (in eukaryotes) moves ahead of the

ATP adenosine triphosphate, a high-energy compound used to power cell processes

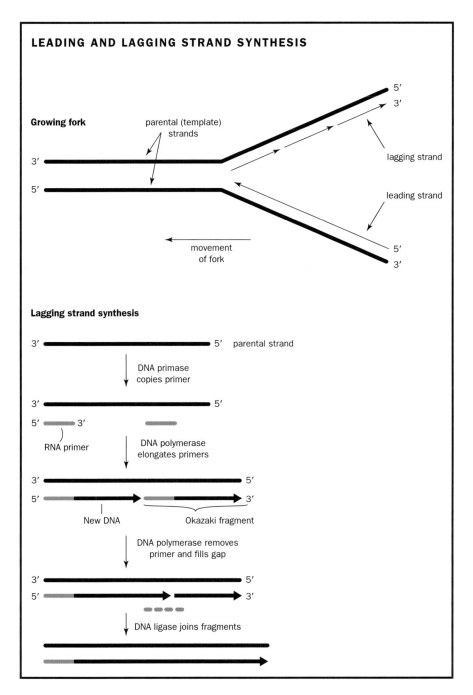

Figure 3. Leading and lagging strand synthesis. Adapted from Lodish, 1999.

replication fork, breaking, swiveling, and rejoining the double helix to relieve the strain.

Leading Strands and Lagging Strands

The coordinated synthesis of the two daughter strands posed an important problem in DNA replication. The two parental strands of DNA run in opposite directions, one from the 5′ to the 3′ end, and the other from the 3′ to the 5′ end. However, all known DNA polymerases catalyze DNA synthesis in only one direction, from the 5′ to the 3′ end, adding nucleotides only to the 3′ end of the growing chain. The daughter strands, if they were both synthesized continuously, would have to be synthesized in opposite direc-

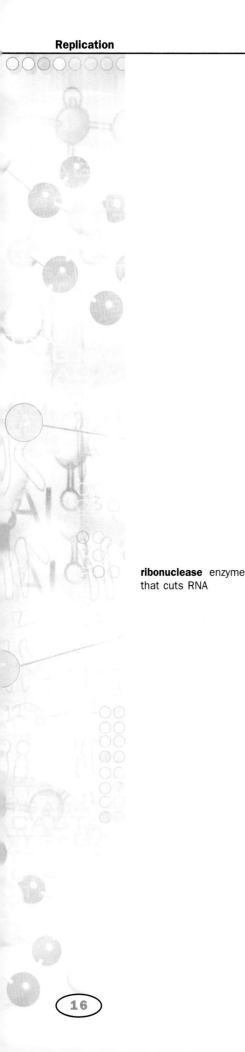

tions, but this is known not to occur. How, then, can the other strand be synthesized?

The resolution of the problem was provided by the demonstration that only one of the two daughter strands, called the leading strand, is synthesized continuously in the overall direction of fork movement, from the 5′ to the 3′ end (see Figure 3). The second daughter strand, called the lagging strand, is made discontinuously in small segments, called Okazaki fragments in honor of their discoverer. Each Okazaki fragment is made in the 5′ to 3′ direction, by a DNA polymerase whose direction of synthesis is backwards compared to the overall direction of fork movement. These fragments are then joined together by an enzyme called DNA ligase.

The Need for Primers

Another property of DNA polymerase poses a second problem in understanding replication. DNA polymerases are unable to initiate synthesis of a new DNA strand from scratch; they can only add nucleotides to the 3′ end of an existing strand, which can be either DNA or RNA. Thus, the synthesis of each strand must be started (primed) by some other enzyme.

The priming problem is solved by a specialized RNA polymerase, called DNA primase, which synthesizes a short (3 to 10 nucleotides) RNA primer strand that DNA polymerase extends. On the leading strand, only one small primer is required at the very beginning. On the lagging strand, however, each Okazaki fragment requires a separate primer.

Before Okazaki fragments can be linked together to form a continuous lagging strand, the RNA primers must be removed and replaced with DNA. In bacteria, this processing is accomplished by the combined action of RNase H and DNA polymerase I. RNase H is a **ribonuclease** that degrades RNA molecules in RNA/DNA double helices. In addition to its polymerase activity, DNA polymerase I is a 5′-to-3′ nuclease, so it too can degrade RNA primers. After the RNA primer is removed and the gap is filled in with the correct DNA, DNA ligase seals the nick between the two Okazaki fragments, making a continuous lagging strand.

ribonuclease enzyme that cuts RNA

DNA Polymerase

The two molecules of DNA polymerase used for the synthesis of both leading and lagging strands in bacteria are both DNA polymerase III. They are actually tethered together at the fork by one of the subunits of the protein, keeping their progress tightly coordinated. Many of the other players involved are also linked, so that the entire complex functions as a large molecular replicating machine.

DNA polymerase III has several special properties that make it suitable for its job. Replication of the leading strand of a bacterial chromosome requires the synthesis of a DNA strand several million bases in length. To prevent the DNA polymerase from "falling off" the template strand during this process, the polymerase has a ring-shaped clamp that encircles and slides along the DNA strand that is being replicated, holding the polymerase in place. This sliding clamp has to be opened like a bracelet in order to be loaded onto the DNA, and the polymerase also contains a special clamp loader that does this job.

A second important property of DNA polymerase III is that it is highly accurate. Any mistakes made in incorporating individual nucleotides cause **mutations**, which are changes in the DNA sequence. These mutations can be harmful to the organism. The accuracy of the DNA polymerase results both from its ability to select the correct nucleotide to incorporate, and from its ability to "proofread" its work.

Appropriate nucleotide selection depends on base-pairing of the incoming nucleotide with the template strand. At this step, the polymerase makes about one mistake per 1,000 to 10,000 incorporations. Following incorporation, the DNA polymerase has a way of checking to see that the nucleotide pairs with the template strand appropriately (that is, A only pairs with T, C only pairs with G). In the event that it does not, the DNA polymerase has a second enzymatic activity, called a proofreading exonuclease, or a 3′-to-5′ exonuclease, that allows it to back up and remove the incorrectly incorporated nucleotide. This ability to proofread reduces the overall error rate to about one error in a million nucleotides incorporated. Other mechanisms detect and remove mismatched base pairs that remain after proofreading and reduce the overall error rate to about one error in a billion.

mutations changes in DNA sequences

Features of Replication in Eukaryotic Cells

The steps in DNA replication in eukaryotic cells are very much the same as the steps in bacterial replication discussed above. The differences in bacterial and eukaryotic replication relate to the details of the proteins that function in each step. Although amino acid sequences of eukaryotic and prokaryotic replication proteins have diverged through evolution, their structures and functions are highly conserved. However, the eukaryotic systems are often somewhat more complicated.

For example, bacteria require only a single DNA polymerase, using DNA polymerase III for both leading and lagging strand synthesis, and are able to survive without DNA polymerase I. In contrast, eukaryotes require at least four DNA polymerases, DNA polymerases α, δ, ϵ, and σ. DNA polymerases δ and ϵ both interact with the sliding clamp, and some evidence suggests that one of these polymerases is used for the leading strand and the other for the lagging strand. One required function of DNA polymerase α is the synthesis of the RNA primers for DNA synthesis. The precise role of DNA polymerase σ is not yet known. A second example is removal of the RNA primers on Okazaki fragments. In eukaryotes, primer removal is carried out by RNase H and two other proteins, Fen1 and Dna2, which replace the 5′-to-3′ exonuclease provided by the bacterial DNA polymerase I in bacteria.

Replication continues until two approaching forks meet. The tips of linear eukaryotic chromosomes, called **telomeres**, require special replication events. Bacterial chromosomes, which contain circular DNA molecules, do not require these special events.

telomeres chromosome tips

Regulating Replication

DNA replication must be tightly coordinated with cell division, so that extra copies of chromosomes are not created and each daughter cell receives exactly the right number of each chromosome. DNA replication is regulated by

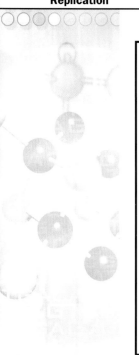

Function(s)	Bacteria	Eukaryotes
single-stranded DNA binding, stimulates DNA polymerase, promotes origin unwinding	SSB (one subunit)	RPA (three subunits)
clamp loader	$\gamma\delta/\delta'\tau$ (5 subunits)	RFC (five subunits)
sliding clamp, holds DNA polymerase on DNA	β (two identical subunits)	PCNA (three identical subunits)
replicative DNA polymerase, proofreading exonuclease	DNA polymerase III	DNA polymerase δ (two subunits) DNA polymerase ε (four subunits)
DNA primase	DnaG	DNA polymerase α (four subunits)
Okazaki fragment processing	DNA polymerase I RNase DNA ligase H	Dna2 FEN1 RNase H DNA ligase I
DNA helicase	DnaB	?
Swivel ahead of replication fork	ω DNA gyrase	Topoisomerase I Topoisomerase II
Initiator protein	DnaA	Origin Recognition Complex (six subunits)

DNA replication proteins.

controlling the assembly of complexes at replication origins. In bacteria, the accumulation of the initiator protein, dnaA, seems to be an important factor in determining when replication begins.

In eukaryotes, DNA replication and cell division are separated by two "gap" cell cycle phases (G_1 and G_2), during which neither DNA replication nor nuclear division occurs. DNA replication occurs during the S (or synthesis) phase, but ORC is thought to bind replication origins throughout the cell cycle. During the G_1 phase of the cell cycle, ORC helps to assemble other replication initiation factors at replication origins to make so-called pre-replicative-complexes (pre-RCs) that are competent to initiate replication during S phase. These other initiation factors include a protein called Cdc6 and a family of six related MCM ("mini-chromosome maintenance") proteins. The functions of these proteins are not yet known; however, the MCM proteins are currently the best candidate for the eukaryotic replicative helicase, and Cdc6 is necessary for MCM proteins to bind DNA. DNA polymerase α also assembles on origins during this time.

Replication initiation is actually triggered at the beginning of S phase by the phosphorylation (addition of a phosphate group to) of one or more proteins in the pre-RC. The enzymes that phosphorylate proteins in the pre-RC are called protein kinases. Once they become active, they not only trigger replication initiation, but they also prevent the assembly of new pre-RCs. Therefore, replication cannot begin again until cells have completed cell division and entered G_1 phase again. SEE ALSO CELL CYCLE; CHROMOSOME, EUKARYOTIC; CHROMOSOME, PROKARYOTIC; DNA; DNA POLYMERASES; DNA REPAIR; MUTATION; NUCLEASES; NUCLEOTIDE; NUCLEUS; TELOMERE.

Carol S. Newlon

Bibliography

Baker, T. A., and S. P. Bell. "Polymerases and the Replisome: Machines within Machines." *Cell* 92 (1998): 295–305.

Cooper, Geoffrey M. *The Cell: A Molecular Approach*. Washington, DC: ASM Press, 1997.

Herendeen, D. R., and T. J. Kelly. "DNA Polymerase III: Running Rings Around the Fork." *Cell* 84 (1996): 5–8.

Lodish, Harvey, et al. *Molecular Cell Biology*, 4th ed. New York: W. H. Freeman, 2000.

Stillman, B. "Cell Cycle Control of DNA Replication." *Science* 274 (1996): 1659–1664.

Internet Resource

Davey, M., and M. O'Donnell. "DNA Replication." Genome Knowledge Base Website. <http://gkb.cshl.org/db/index>.

Reproductive Technology

Successful pregnancy requires ovulation (when an ovary releases an egg into a **fallopian tube**), transport of the egg partway down the fallopian tube, movement of sperm from the vagina to the fallopian tube, penetration by the sperm of the egg's protective layer, and implantation of the fertilized egg in the uterus.

fallopian tube the tube through which eggs pass to the uterus

In the United States, infertility is an issue of great concern to many couples of childbearing age. More than 15 percent of all such couples are estimated to be infertile. In a 1995 study by the Centers for Disease Control and Prevention, 10 percent of 10,847 women interviewed, a percentage that represents 6.1 million women of childbearing age nationwide, reported having experienced some problems getting pregnant or carrying a baby to term. Of this group, about half were fertile themselves but had infertile partners. The number of women seeking professional assistance to deal with infertility problems is increasing every year (600,000 in 1968, 1.35 million in 1988, 2.7 million in 1995), and it is reasonable to believe that this trend will continue unabated well into the twenty-first century.

Pregnancy and Infertility

There are many causes of infertility. Abnormal semen causes the infertility problems of about 30 percent of couples seeking treatment. Tubal disease and **endometriosis** in the female partner account for another 30 percent. A female partner's failure to ovulate accounts for 15 percent, and the inability of sperm to penetrate the woman's cervical mucus accounts for another 10 percent. The final 15 percent of couples seeking treatment are infertile for reasons that cannot be diagnosed.

endometriosis disorder of the endometrium, the lining of the uterus

Many couples can be helped to overcome infertility through hormonal or surgical interventions. Women experiencing ovulation disorders may benefit from treatment with oral drugs such as clomiphene citrate, or through the injection of gonadotropins, such as follicle-stimulating hormone, which has about a 75 percent success rate. Women with tubal disease can be helped by various types of reconstructive surgery, but the success rate is only about 33 percent.

However, many infertile couples cannot be helped by such standard methods of treatment. Instead, as a last resort, couples that want children must turn to newer techniques that bypass one or more steps in the usual physiological processes of ovulation, fertilization, and implantation. Commonly

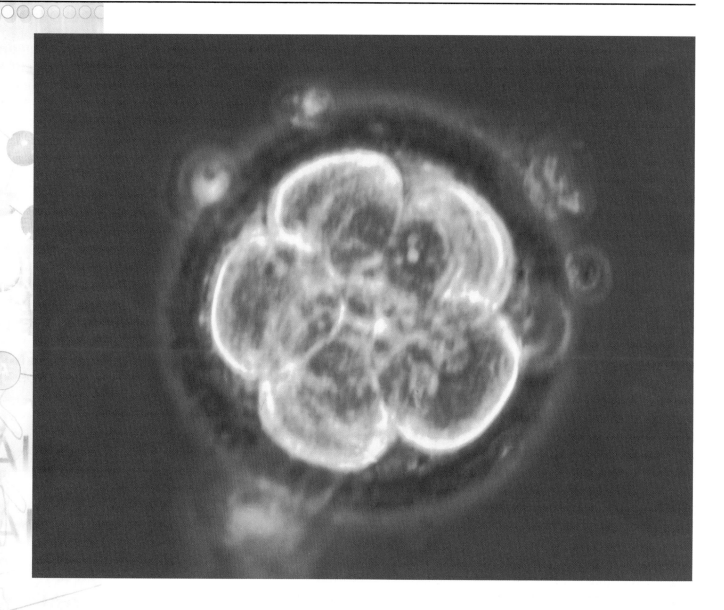

This eight-cell human embryo is an example of a blastocyst.

in vitro "in glass"; in lab apparatus, rather than within a living organism

cryopreservation use of very cold temperatures to preserve a sample

referred to as "assisted reproductive technology," these techniques include **in vitro** fertilization (IVF), gamete intrafallopian transfer, zygote intrafallopian transfer, donor insemination, egg donation, embryo **cryopreservation**, intra-cytoplasmic sperm injection, tubal embryo stage transfer, and intrauterine insemination.

In vitro Fertilization

When performed by an experienced practitioner and in an experienced clinic, IVF generally results in pregnancy rates of about 28 percent after one attempt and 51 percent after three. One study has reported the pregnancy rate after six attempts as being 56 percent. Another has reported it as being 66 percent.

Generally, one attempt at IVF is made per menstrual cycle. The IVF process begins when couples are first screened. Clinicians first must rule out infertility in the male partner. If the problem is with the female partner, various courses of treatment may be available. Generally, couples are expected to try to achieve pregnancy for a year after the initial screening before intervention is attempted. However, if a woman is in her late thirties or older, or if she is experiencing irregular menstruation, a clinical investigation may begin earlier.

Especially in older women, the blood level of follicle-stimulating hormone, a hormone that acts on the ovary to stimulate the maturation of viable eggs, is measured. If the hormone's level is found to be elevated early in a woman's menstrual cycle (after the first week of the new cycle), her ovaries may not be responding to it. In that case, hormonal treatment to stimulate ovulation would be ineffective, and assisted reproductive technology would be unable to help achieve pregnancy. Elevated estrogen levels at day three would also indicate that the ovaries are not responding correctly to estrogen or hormones.

In women whose ovaries are capable of generating viable eggs, the first step in IVF is referred to as "superovulation." To increase the chance of success, the woman's ovaries are stimulated to develop many follicles. Normally, only one or perhaps two follicles develop and are released by an ovary during a single menstrual cycle, which is why usually only one or, on rare occasions, two children are born. In superovulation, a doctor forces multiple follicles to develop so that many **oocytes** can be collected.

oocytes egg cells

To stimulate the ovaries to develop many follicles, the woman undergoes the "long protocol." The action of the pituitary gland is suppressed hormonally, and ten days later the woman is treated with follicle-stimulating hormone. To see how well her ovaries are responding to the hormone, doctors measure estrogen blood levels and observe the ovaries with ultrasound scans. The number and size of the follicles can be visualized. When the doctors judge that the time is right (that is, when the follicle is enlarged to the point that it protrudes above the surface of the ovary), they give the woman human chorionic gonadotropin, wait thirty-six hours, and collect the oocytes from the mature follicles.

In the past, to collect follicles, doctors performed laparoscopy, in which a thin optical tube with a light (called a laproscope) is inserted through a very small incision in the abdominal wall, and the pelvic organs are viewed with fiber optics. Today, the use of a needle guided by ultrasound makes the procedure much faster. The ovary is visualized, mature follicles are located, the needle is inserted, and the follicular fluid that contains the mature oocyte (the unreleased egg) is **aspirated**. The doctors may collect up to eleven oocytes from a single patient.

aspirated removed with a needle and syringe

Viable sperm are collected from the man and washed in a special solution that activates them so they can fertilize the egg. The process of sperm activation is called "capacitation" and normally occurs when sperm are ejaculated and enter the female reproductive tract. Capacitation involves activating enzymes in the sperm's **acrosomal cap**, allowing the sperm head, which contains the sperm's genetic material, to penetrate the outer and inner membranes of the egg (zona pellucida and vitelline membrane). For males with azoospermia, microsurgical or aspiration techniques can directly extract sperm from either the **epididymis** or the testicles. Azoospermia is the most severe form of male infertility, caused by obstructions in the genital tract or by testicular failure.

acrosomal cap tip of sperm cell that contains digestive enzymes for penetrating the egg

epididymis tube above the testes for storage and maturation of sperm

To allow fertilization to take place, a single egg and about 100,000 sperm are placed together in special culture medium and incubated for about

twenty-four hours. Doctors then use a microscope to see if there are two pronuclei (one from the egg and one from the sperm) in the egg, indicating that fertilization occurred. In some cases, the sperm are unable to penetrate the egg. They may be unable to swim correctly, or they may not have been capacitated successfully. In the past, the only solution was to use sperm from another man. Now, however, sperm can be injected directly into the egg's **cytoplasm**, in a process called microassisted fertilization.

cytoplasm the material in a cell, excluding the nucleus

There are three ways used successfully by doctors and researchers to micro-fertilize the egg. The first is "zona drilling," in which a hole is punched into the zona pellucida, letting sperm penetrate the egg. The second method is called "subzonal sperm insertion," in which a sperm is injected directly under the zona pellucida. A third, related method is "intracytoplasmic sperm injection," in which a sperm is injected directly into the egg cytoplasm. The second method is reported to have a higher fertilization success rate (59%) than the third (13%).

The Risks of IVF

In some women, the drugs used to promote superovulation may cause side effects, including mood swings. Some investigators have suggested that procedures used in assisted reproductive technology may not be safe because of the potential for increased ovarian hyperstimulation syndrome and bone loss from the hormone treatments. Ovarian hyperstimulation syndrome occurs at mild levels in 23 percent of women undergoing the treatments, at moderate levels in 3 percent, and at severe levels in 0.1 percent. Complications also may arise as a result of the surgical procedures involved in egg retrieval and embryo transfer. Such complications include pelvic and other infections, which occur in 0.15 percent to 1.2 percent of women, complications from anesthesia, which occur in 0.2 percent, and internal injuries, which occur in 0.38 percent. Although the incidence of such complications is low, every chemical or surgical intervention is associated with risks, and potential patients should be aware of this.

Other concerns regarding the long-term effects of assisted reproductive technology include the increased incidence of spontaneous abortions, which occur in 20 percent of women, and ectopic and heterotopic pregnancies, which occur in 5.5 and 1.2 percent of women, respectively. In heterotopic pregnancies, an embryo is implanted outside the uterus.

There have been conflicting reports as to whether there is a link between the use of fertility drugs and ovarian cancer. Overall, results from many studies do not seem to support such a connection. Some researchers have suggested that the use of the drugs may increase the risk of ovarian cancer later in life, but this is difficult to prove or disprove because the techniques have not been around long enough to assess long-term effects. There have been no reports of any increase in abnormalities in children born using micro-assisted fertilization, though critics have questioned whether the techniques might increase the incidence of such abnormalities.

Embryo Transfer Techniques

zygote fertilized egg

media nutrient source

Once the egg is fertilized, the two pronuclei fuse to form the early embryo, or "**zygote**." The zygotes are placed in culture **media**, where they undergo cell division, cleaving repeatedly and passing through the two-cell, four-cell, and eight-cell stages. Within seventy-two hours, the zygote develops into the morula, the solid mass of blastomeres formed by the cleavage of the fertilized ovum (egg). After about three to five days in culture, the zygote has become a hollow ball of cells called the **blastocyst**. During normal embryogenesis, it is the blastocyst that is implanted in the endometrial lining of the uterus.

blastocyst early stage of embryonic development

While the embryos are in culture, problems in their development may become apparent. After embryos with evident problems are discarded, one or more cultured embryos are transferred into the uterus, where, it is hoped, one will become implanted and develop into a healthy, full-term baby. Embryo transfer replaces the natural process in which the embryo passes down the fallopian tube and into the uterus, prior to implantation.

Although this transfer is relatively simple and often takes only a few minutes, the rate of successful implantation is low. Usually, two embryos are transferred, but still only one of five women become pregnant. Most doctors who perform IVF procedures adhere to the limit of two embryos per woman to minimize the risk of multiple pregnancies. In most cases, three embryos are transferred in women older than thirty-five. In the United Kingdom, it is illegal for an IVF doctor to transfer more than three embryos at a time into a woman.

A number of factors play a role in whether embryo transfer leads to a baby being born. Success rate is higher if embryo transfer takes place between forty-eight and seventy-two hours after oocyte collection. When more than one embryo is transferred at the same time, the success rate increases, but so does the chance for multiple pregnancies. As noted above, the maximum number transferred should never exceed three. Probably the single most important factor determining whether or not a successful embryo implantation will take place is the donated egg's age. Embryos formed from eggs donated by younger women have a higher implantation success rate than do embryos formed from eggs donated by older women. The age of the host uterus appears to have little or no effect on outcome.

Gamete Intrafallopian Transfer

An alternative to IVF and intrauterine embryo transfer is gamete intrafallopian transfer (GIFT), introduced more than twenty years ago. In this procedure, the egg and sperm are collected as they would be for IVF procedures. However, instead of allowing fertilization to take place in a culture dish, the egg and sperm are transferred surgically into the woman's fallopian tube. This allows fertilization to occur in the fallopian tube, just as occurs in a natural pregnancy. The transfer can only be performed in women with healthy and functional fallopian tubes, and the sperm used for fertilization must be completely normal and capable of swimming.

After the transfer is made, doctors have no way of knowing if normal fertilization actually takes place until an embryo has implanted in the uterine wall. However, the procedure's success rate (35%) is higher than the success rate for IVF. Another related technique is zygote intrafallopian transfer, in which the egg is fertilized in vitro and the zygote is transferred surgically into the fallopian tube. Other techniques include tubal embryo transfer, in which an embryo already undergoing cleavage is transferred.

Intrauterine Insemination

Intrauterine insemination is used when a couple's inability to conceive a child is caused by the sperm's inability to reach the egg. Sperm must move through the uterus and enter the fallopian tube before they can fertilize the egg. Anything that prevents the sperm from making this trip will block conception. Coital or ejaculatory disorders can limit the sperm's travels, sperm antibodies in the female reproductive tract can kill the sperm, and sperm may be unable to penetrate the cervical mucus.

To help the sperm reach the egg, the female is treated with human chorionic gonadotropin to induce multiple ovulation. The number of follicles that are induced is monitored by ultrasound. Washed sperm from the male partner are injected through the cervical opening, into the uterus. The pregnancy rate using this procedure is about 10 percent.

Donor Insemination and Egg Donation

Donor insemination is used when sperm are incapable of fertilizing the egg. Usually this occurs if the male produces very little or no sperm. Sometimes, donor sperm is used when the male partner is the carrier of a genetic disorder that could be transmitted to the baby. Sperm donors should be between ages eighteen and fifty-five, and all should be screened for genetic disorders, such as cystic fibrosis, and for various types of chromosomal abnormalities and infectious disease, including hepatitis, syphilis, cytomegalovirus, and HIV. As with the use of intrauterine insemination, the female partner undergoes ovarian stimulation to maximize the number of follicles released during ovulation. Pregnancy rates resulting from the use of donor insemination are between 32 percent and 50 percent after ten inseminations.

As with donor insemination, egg donation is used when the woman cannot ovulate or is the carrier of a genetic disease. Egg donors must be younger than thirty-five years and must be screened for the same set of conditions as sperm donors. Donors are treated with drugs to stimulate ovulation, after which the eggs are fertilized with the sperm from the male partner and the embryos are transferred to the uterus of the female partner (other procedures can also be used). Growth and development of the embryos then follow the natural processes.

Surrogacy and Cryopreservation

Surrogacy, in which pregnancy occurs in another woman, can supply a couple with an alternative if the woman partner cannot carry the baby to term in her own uterus. In some cases, if the woman cannot supply the egg, sperm from the male partner can be used to inseminate the surrogate mother, who carries the baby to term. Alternatively, if the female partner can produce her own egg, sperm from the male partner can be used to fertilize the egg, and the resulting pre-embryo can be transferred to the uterus of the surrogate mother to grow and develop. Legal controversies resulting from these arrangements have become common in the last few years, so the arrangements should be carefully reviewed by all parties, along with experts in the field, before any final decisions are made.

Frozen sperm and embryos effectively retain their viability for many years. The use of frozen human blastocysts is associated with a 10 percent successful pregnancy rate. Oocyte freezing is much less successful, possibly because oocytes may be genetically damaged or killed in the freezing and thawing. Embryos produced from such cryopreserved eggs have a high incidence of **aneuploidy**, and they are slow to cleave and develop even if they appear to be genetically undamaged. Various research groups are trying to solve this problem.

aneuploidy abnormal chromosome numbers

Age as a Factor

Age must be taken into account when couples are considering assisted reproductive technology. In humans, the age of the oocyte, not the age of the uterus, is the main cause of reproductive failure in IVF and embryo transfer techniques. Embryos formed from older oocytes demonstrate an increased incidence of aneuploidy. In some other species, such as in rabbits, an aging uterus can keep an embryo from being implanted. The use of cryopreservation to circumvent reproductive failure in humans, cattle, and horses has already been successfully employed and is likely to be developed further.

Sperm generated by older men are capable of successfully producing normal embryos. However, as sperm age, they are exponentially more likely to contain new gene **mutations**. Older oocytes, on the other hand, do not appear to be more likely to contain new mutations. Scientists are unsure exactly how age affects oocyte integrity. Oocyte maturation takes place only before birth in the female, so no new oocytes are produced during the entire reproductive life of the female. This is quite different from spermatogenesis, which can continue into old age. Thus, oocytes from older women may be forty or more years old when they are collected and used to form the embryos.

mutations changes in DNA sequences

Oocytes must reach full maturity before they can be ovulated normally and before they can be fertilized, even artificially, to form embryos. If immature oocytes could be artificially forced to mature in vitro, follicles could be taken from the ovaries of dying or dead women, or from cancer patients planning on undergoing chemotherapy treatments, which can damage oocytes. Unlike immature oocytes, immature sperm can effectively be used in fertilization. Additional research is needed in this area of assisted reproductive technology.

Legal, Ethical, and Moral Considerations

The use of these powerful techniques to facilitate reproduction in both humans and animals (the techniques can be used in cattle and pigs, and in the conservation of endangered wildlife) must be balanced against legal, ethical, and moral concerns. For example, would it be permissible to revive extinct animal species? Although a Jurassic Park-like scenario to reanimate extinct dinosaurs is not scientifically credible at this time, what if it became possible to use this technology to form embryos and clone an extinct mammoth, or the passenger pigeon? And what if we can do this for extinct humans? Just because we can develop the capability, would it be acceptable? What are the ethics involved?

Other concerns include questions about how long embryos should remain frozen and who owns frozen embryos not used by the parents. What happens if the parents separate, divorce, or die? What about the legal entanglements involved with surrogacy? Already in the media there have been a number of such cases reported. With the expected increase of these procedures in the future, it is likely that such complex questions will only escalate. Finally, there are basic concerns about helping people sidestep the natural birth process to bring into the world a new human. SEE ALSO CELL CYCLE; CHROMOSOMAL ABERRATIONS; CLONING ORGANISMS; FERTILIZATION; LEGAL ISSUES; MEIOSIS; PRENATAL DIAGNOSIS; REPRODUCTIVE TECHNOLOGY: ETHICAL ISSUES.

Charles J. Grossman and Robert Baumiller

Bibliography

Ryan, Michael. "Countdown to a Baby." *New Yorker* (July 21, 2002): 68–77.

Schultz, Richard M., and Carmen J. Williams. "The Science of ART." *Science* 21 (June 2002): 2188–2190.

Reproductive Technology: Ethical Issues

Reproductive technology encompasses a range of techniques used to overcome infertility, increase fertility, influence or choose the genetic characteristics of offspring, or alter the characteristics of a population. Each type of reproductive technology brings with it a range of ethical issues. With the accelerated pace of progress in modern medical technology, these issues have been brought squarely into the public arena, where they continue to provoke controversies involving the boundaries of government control, private choice, religious belief, and parental wishes.

Recoiling from Eugenics

Humans' ability to selectively breed desired characteristics into domestic animals and plants, combined with pride and concern for family and for national and ethnic heritage, has led historically to multiple suggestions and experiments aimed at "improving" the human race. These ideas gave rise to the **eugenics** movements in Great Britain and in the United States in the late nineteenth and early twentieth century, preceding and influencing the Nazi "racial hygiene" experiments of the 1930s and 1940s. The horror of these experiments caused an almost universal backlash against eugenic programs that continues to this day.

Nevertheless, from time to time suggestions are made to improve the human gene pool so as to produce people genetically suited for specific tasks, or to improve the general levels of intelligence or other traits that are considered desirable. Numerous ethicists have made strong arguments against such suggestions. The novelist Aldous Huxley decried them in his famous satire *Brave New World*. Paul Ramsey, a Methodist moralist and one of the founders of modern bioethics, argued in his book *Fabricated Man* that any change from natural procreation to mechanized reproductive technology would be harmful to individuals and to society in general.

eugenics movement to "improve" the gene pool by selective breeding

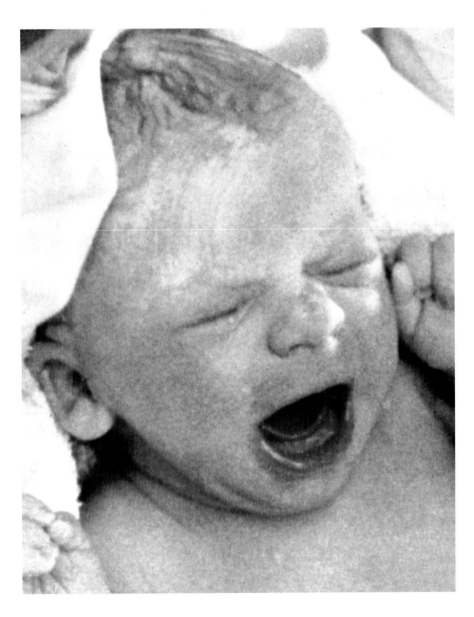

Louise Joy Brown, the first baby conceived by in vitro fertilization was born in England on July 25, 1978.

A child is generally seen as a kind of gift of nature, conceived out of the love and passion of two people. However, as more technology is used in childbirth, there is a concern that the child may be seen increasingly as a commodity whose major purpose is to satisfy the emotional needs of the parents.

Another twenty or thirty years must pass before data will be available on the well-being of children born to couples as a result of reproductive technology, although anecdotal reports are positive. Data about children who have been brought into the world by technology for reasons other than infertility may have to wait another fifty to one hundred years before being analyzed because such children are few in number and difficult to find and study.

The Poles of the Debate

Because reproductive technology encompasses a wide variety of both goals and techniques, there is no single ethical position held by most thinkers considering its ethics. For many, the central, underlying question is when "personhood" begins, in the life of a fertilized **ovum**.

ovum egg

27

The most conservative stance, opposing all interventions in natural procreation and therefore opposing most reproductive methodologies, is adopted by the Roman Catholic Church and a number of other conservative groups, all of whom invoke some variation of Natural Law ethics. The conservative basis of this nonintervention stance is based on granting full respect to the conceptus from the moment of conception, defined as when a sperm penetrates an ovum.

Some methods used to overcome obstacles to childbirth are permitted by many of these groups. Many would not oppose concentrating a sperm sample from a husband whose sperm count was low, nor would they oppose bringing an ovum around a blocked fallopian tube so sperm could reach the ovum within the mother's reproductive tract. They also would not oppose administering medication to enhance ovulation. Each of these methods involves manipulating the sperm or ovum before conception.

Others who assign the beginning of personhood to later stages of development advocate a greater freedom to utilize reproductive technology depending on the ethical merits of specific situations and on the rights of other individuals who are involved. Within this group, a central question is where to draw the boundary between parental freedoms to choose methods of procreation and to influence the characteristics of their child on the one hand, and, on the other, societal interests in protecting the unborn.

Donor Gametes

One of the oldest and least controversial reproductive technologies is the use of donor sperm to overcome low sperm count on the part of the male or to avoid inheritance of some genetic condition that the male might pass on to his child. Donation is usually anonymous, but some characteristics of the donor are known.

Ethical issues arising in sperm donation include the extent to which parents have the right to choose desirable characteristics in the genetic father of the child, and the right of the child to eventually learn the identity of the father. Each of these issues has precedent in nonassisted reproduction, since prospective parents do choose their mates, and anonymous parenthood occurs in many adoptions. Payment for the sperm sample is generally low enough that the incentive to donate is not thought to be coercive for the donor, and so is not a significant ethical issue.

in vitro "in glass"; in lab apparatus, rather than within a living organism

Donor ova (eggs) are now sometimes used in combination with **in vitro** fertilization technology. Women who are unable to ovulate, or whose ova might transmit a genetic condition they do not want to pass on, use this method. Ovum donation poses risks to the donor. Medication usually is given to cause the release of excess ova, and laparoscopy (in which a needle is inserted through the abdomen) is used to retrieve the ova. Reimbursement is higher and can easily become coercive. Large sums of money have been offered to young women at prestigious colleges, and glamorous potential donors have asked for as much as $100,000.

In Vitro Fertilization and Surrogacy

In vitro fertilization, even with anonymously donated ovum or sperm, is usually accepted as enabling a couple to experience gestation and create a family. However, the in vitro procedure typically creates more embryos than will be implanted. These excess embryos are usually frozen and may remain viable (able to develop normally) for several years. Is it ethical to create embryos that will never be developed? Are the embryos of a couple joint property, which can be divided upon divorce, or should they be considered children, with custody awarded to one parent or the other? What should be done with frozen embryos as they lose their viability?

The ability to fertilize an ovum in a glass dish makes it possible to combine gametes from any man and woman and implant the fertilized ovum into any other woman. Should this be done for other than a husband and wife? For instance, should a reproductive specialist be allowed to refuse to help an unmarried or same-sex couple have a child? Antidiscrimination laws cover some of these issues, but the application of this law to reproductive technology services is still developing.

In vitro fertilization also raises the possibility of employing a woman to be a surrogate mother, acting as the carrier of a child who is expected to be raised by another woman. Women have contracted to render this service in exchange for having all their medical costs covered and in exchange for a stipend to be paid when the baby is delivered. The ethical and legal problems posed by such arrangements have drawn much attention, and many states regulate the boundaries of such agreements.

The birth mother (surrogate) usually signs a contract to allow the adoption of the baby upon birth. However, hormonal changes during pregnancy, especially the increased levels of oxytocin, can produce a variety of behavioral changes, often termed "maternal instinct," in the pregnant woman. The woman may have been quite willing to surrender the child upon birth, but she may have become unwilling to do so after birth.

In general, the birth mother has the principal right to decide about fulfilling her contract to give over the baby, but the state has stepped in, making contrary decisions when the welfare of the child is at stake. Surrogacy has other potential problems, such as what happens when there are prenatally diagnosed abnormalities and termination decisions must be made or when there are differing standards of prenatal care between the surrogate and the contracting party, and when there is a desire for control of the contracting individual or couple over the surrogate.

Genetic Selection and Medical Motives

Ova and sperm can be genotyped so that particular genes or combinations of genes can be selected. While this is usually very expensive and not yet foolproof, success has been reported in influencing the sex of a child.

Cells from early embryos can be removed for even more precise genotypic determination and selection. Such selection has the potential of being more and more precise. Technology may even become capable of altering traits, as the **genome** becomes better understood. What kind of genetic selection should be allowable in embryos? Should one avoid mental retardation in potential children? Should one select for possibilities of greater intelligence? Is it ethical to choose to have a child only of a particular sex?

Genetic analysis also allows parents to have children whose genes can help others. But should a child be brought into the world simply for the

genome the total genetic material in a cell or organism

sake of another, or should it be brought in only for its own sake? In a widely reported case in 2000, a couple selected for implantation an embryo that was found to be a perfect bone marrow match for their child, whose own marrow was failing due to Fanconi anemia.

Testing showed that the embryo did not have the recessive disease and that his marrow would not be rejected by his sister, who needed a transplant. The parents attested to their desire for another child, they refused to give birth prematurely for the sick child's sake and they refused to put the infant in any risk by extracting marrow for transplant. They used only the stem cells from the placenta to begin the repopulation of the older daughter's marrow. The process worked! This case highlights the difficult issues involved, and probably represents the best process and outcome possible in such a case. The future may bring more problematic cases as the technology advances to allow treatment of more conditions through tissue or **stem cell** transplantation.

stem cell cell capable of differentiating into multiple other cell types

Stem Cells

The existence of stem cells—cells with the potential to develop into a wide variety of other cell types—presents other ethical issues. Placental stem cells may not be able to become every type of cell in the body. In contrast, stem cells derived from developing embryos can. People who believe abortion is ethically acceptable usually have little problem with stem cell collection, because they do not accept the embryo as a person at this stage of gestation.

Most people who believe that personhood begins at conception object to harvesting stem cells from embryos, even from the many thousands of embryos frozen in reproductive clinics that gradually lose their viability as years pass. In 2001 President George W. Bush declared that federal funds could be used for research only on stem cell lines that were already in existence, and that the creation of new lines could not be funded by taxpayer money. Stem cell research is allowed if funded by private sources. Many scientists feel federal money should be allowed to fund stem cell research, citing the enormous potential benefits it can bring. SEE ALSO CLONING ORGANISMS; EUGENICS; GENETIC TESTING: LEGAL ISSUES; PRENATAL DIAGNOSIS; PRIVACY; REPRODUCTIVE TECHNOLOGY.

Robert C. Baumiller and Charles J. Grossman

Bibliography

Andrews, Lori B. *The Clone Age: Adventures in the New World of Reproductive Technology.* New York: Henry Holt, 2000.

Ramsey, Paul. *Fabricated Man.* New Haven, CT: Yale University Press, 1970.

Sacred Congregation for the Doctrine of the Faith 1987. "Introduction on Respect for Human Life in Its Origin and on the Dignity of Procreation." *Origins* 16, no. 40 (March 19, 1987): 698–711.

Internet Resource

Ratzinger, Joseph, and Alberto Bovone. "Respect for Human Life: Congregation for the Doctrine of the Faith." Rome. The Feast of the Chair of St. Peter, the Apostle. February 1987. <http://www.ewtn.com/library/curia/cdfhuman.htm>.

Restriction Enzymes

Restriction enzymes are bacterial proteins that recognize specific DNA sequences and cut DNA at or near the recognition site. These enzymes are widely used in molecular genetics for analyzing DNA and creating **recombinant DNA** molecules.

Biological Function and Historical Background

Restriction enzymes apparently evolved as a primitive immune system in bacteria. If viruses enter a bacterial cell containing restriction enzymes, the viral DNA is fragmented. Destruction of the viral DNA prevents destruction of the bacterial cell by the virus. The term "restriction" derives from the phenomenon in which bacterial viruses are restricted from replicating in certain strains of bacteria by enzymes that cleave the viral DNA, but leave the bacterial DNA untouched. In bacteria, restriction enzymes form a system with modification enzymes that **methylate** the bacterial DNA. Methylation of DNA at the recognition sequence typically protects the microbe from cleaving its own DNA.

Since the 1970s, restriction enzymes have had a very important role in recombinant DNA techniques, in both the creation and analysis of recombinant DNA molecules. The first restriction enzyme was isolated and characterized in 1968, and over 3,400 restriction enzymes have been discovered since. Of these enzymes, over 540 are currently commercially available.

Nomenclature and Classification

Restriction enzymes are named based on the organism in which they were discovered. For example, the enzyme *Hind* III was isolated from *Haemophilus influenzae*, strain Rd. The first three letters of the name are italicized because they abbreviate the genus and species names of the organism. The fourth letter typically comes from the bacterial strain designation. The Roman numerals are used to identify specific enzymes from bacteria that contain multiple restriction enzymes. Typically, the Roman numeral indicates the order in which restriction enzymes were discovered in a particular strain.

There are three classes of restriction enzymes, labeled types I, II, and III. Type I restriction systems consist of a single enzyme that performs both modification (methylation) and restriction activities. These enzymes recognize specific DNA sequences, but cleave the DNA strand randomly, at least 1,000 **base pairs** (bp) away from the recognition site. Type III restriction systems have separate enzymes for restriction and methylation, but these enzymes share a common subunit. These enzymes recognize specific DNA sequences, but cleave DNA at random sequences approximately twenty-five bp from the recognition sequence. Neither type I nor type III restriction systems have found much application in recombinant DNA techniques.

Type II restriction enzymes, in contrast, are heavily used in recombinant DNA techniques. Type II enzymes consist of single, separate proteins for restriction and modification. One enzyme recognizes and cuts DNA, the other enzyme recognizes and methylates the DNA. Type II restriction enzymes cleave the DNA sequence at the same site at which they recognize it. The only exception are type IIs (shifted) restriction enzymes, which cleave

recombinant DNA DNA formed by combining segments of DNA, usually from different types of organisms

methylate add a methyl group

base pairs two nucleotides (either DNA or RNA) linked by weak bonds

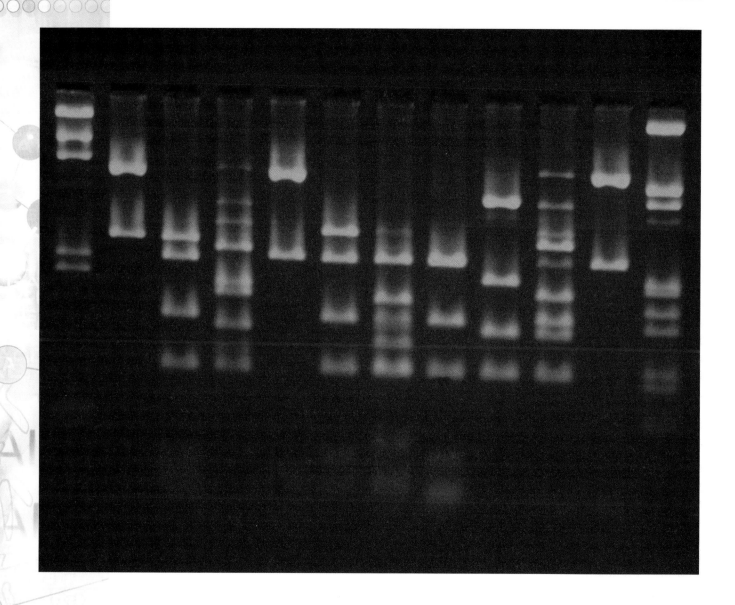

Different restriction enzymes cleave DNA at specific recognition sequences. Cleaving a single piece of DNA with multiple restriction enzymes creates a "DNA fingerprint." The pattern of fragments can be compared to similar DNA from another source treated with the same enzymes, to determine if the two are identical or different.

DNA on one side of the recognition sequence, within twenty nucleotides of the recognition site. Type II restriction enzymes discovered to date collectively recognize over 200 different DNA sequences.

Type II restriction enzymes can cleave DNA in one of three possible ways. In one case, these enzymes cleave both DNA strands in the middle of a recognition sequence, generating blunt ends. For example:

Sma I 5′- CCCGGG- 3′   5′- CCC GGG-3′
3′- GGGCCC- 5′   3′- GGG CCC-5′

(The notations 5′ and 3′ are used to indicate the orientation of a DNA molecule. The numbers 5 and 3 refer to specific carbon atoms in the deoxyribose sugar in DNA.)

These blunt ended fragments can be joined to any other DNA fragment with blunt ends, making these enzymes useful for certain types of DNA cloning experiments.

Type II restriction enzymes can also cleave DNA to leave a 3′ ("three prime") overhang. (An overhang means that the restriction enzyme leaves a short single-stranded "tail" of DNA at the site where the DNA was cut.) These 3′ overhanging ends can only join to another compatible 3′ overhanging end (that is, an end with the same sequence in the overhang).

```
              ▼
Pst I   5′- CTGCAG- 3′    -CTGCA-3′           G-
        3′- GACGTC- 5′    G            3′-ACGTC
              ▲
```

Finally, some type II enzymes can generate 5′ overhanging DNA ends, which can only be joined to a compatible 5′ end.

```
             ▼
Bam HI  5′- GGATCC-3′          G           5′-GATCC
        3′- CCTAGG-3′          CCTAG-5′           G
             ▲
```

In the type II restriction enzymes discovered to date, the recognition sequences range from 4 bp to 9 bp long. Cleavage will not occur unless the full length of the recognition sequence is encountered. Enzymes with a short recognition sequence cut DNA frequently; restriction enzymes with 8 or 9 bp sequences typically cut DNA very infrequently, because these longer sequences are less common in the target DNA.

Use of Restriction Enzymes in Biotechnology

The ability of restriction enzymes to reproducibly cut DNA at specific sequences has led to the widespread use of these tools in many molecular genetics techniques. Restriction enzymes can be used to map DNA fragments or genomes. Mapping means determining the order of the restriction enzyme sites in the genome. These maps form a foundation for much other genetic analysis. Restriction enzymes are also frequently used to verify the identity of a specific DNA fragment, based on the known restriction enzyme sites that it contains.

Perhaps the most important use of restriction enzymes has been in the generation of recombinant DNA molecules, which are DNAs that consist of genes or DNA fragments from two different organisms. Typically, bacterial DNA in the form of a plasmid (a small, circular DNA molecule) is joined to another piece of DNA (a gene) from another organism of interest. Restriction enzymes are used at several points in this process. They are used to digest the DNA from the experimental organism, in order to prepare the DNA for cloning. Then a bacterial plasmid or bacterial virus is digested with an enzyme that yields compatible ends. These compatible ends could be blunt (no overhang), or have complementary overhanging sequences. DNA from the experimental organism is mixed with DNA from the plasmid or virus, and the DNAs are joined with an enzyme called DNA **ligase**. As noted above, the identity of the recombinant DNA molecule is often verified by restriction enzyme digestion.

Restriction enzymes also have applications in several methods for identifying individuals or strains of a particular species. Pulsed field **gel electrophoresis** is a technique for separating large DNA fragments, typically fragments resulting from digesting a bacterial genome with a rare-cutting restriction enzyme. The reproducible pattern of DNA bands that is pro-

ligase enzyme that repairs breaks in DNA

gel electrophoresis technique for separation of molecules based on size and charge

duced can be used to distinguish different strains of bacteria, and help pinpoint if a particular strain was the cause of a widespread disease outbreak, for example.

Restriction fragment length polymorphism (RFLP) analysis has been widely used for identification of individuals (humans and other species). In this technique, genomic DNA is isolated, digested with a restriction enzyme, separated by size in an agarose gel, then transferred to a membrane. The digested DNA on the membrane is allowed to bind to a radioactively or fluorescently labeled probe that targets specific sequences that are bracketed by restriction enzyme sites. The size of these fragments varies in different individuals, generating a "biological bar code" of restriction enzyme-digested DNA fragments, a pattern that is unique to each individual.

Restriction enzymes are likely to remain an important tool in modern genetics. The reproducibility of restriction enzyme digestion has made these enzymes critical components of many important recombinant DNA techniques. SEE ALSO Biotechnology; Cloning Genes; Gel Electrophoresis; Mapping; Methylation; Nucleases; Polymorphisms; Recombinant DNA.

Patrick G. Guilfoile

Bibliography

Bloom, Mark V., Greg A. Freyer, and David A. Micklos. *Laboratory DNA Science: An Introduction to Recombinant DNA Techniques and Methods of Genome Analysis.* Menlo Park, CA: Addison-Wesley, 1996.

Cooper, Geoffrey. *The Cell: A Molecular Approach.* Washington, DC: ASM Press, 1997.

Kreuzer, Helen, and Adrianne Massey. *Recombinant DNA and Biotechnology*, 2nd ed. Washington, DC: ASM Press, 2000.

Lodish, Harvey, et al. *Molecular Cell Biology*, 4th ed. New York: W. H. Freeman, 2000.

Old, R. W., and S. B. Primrose. *Principles of Gene Manipulation*, 5th ed. London: Blackwell Scientific Publications, 1994.

Internet Resource

Roberts, Richard J., and Dana Macelis. *Rebase.* <http://rebase.neb.com>.

Retrovirus

Retroviruses are RNA-containing viruses that use the enzyme reverse transcriptase to copy their RNA into the DNA of a host cell. Retroviruses have been isolated from a variety of vertebrate species, including humans, other mammals, reptiles, and fish. The family Retroviridae includes such important human **pathogens** as human immunodeficiency virus (HIV) and human Tlymphotropic virus (HTLV), the causes of AIDS and adult T-cell leukemia respectively. The study of this virus family has led to the discovery of **oncogenes**, resulting in a quantum advance in the field of cancer genetics. Retroviruses are also valuable research tools in molecular biology and gene therapy.

pathogens disease-causing organisms

oncogenes genes that cause cancer

Characteristics

The classification of retroviruses is based on comparisons of the size of the genome and morphologic characteristics (see Table 1). The genomic RNA

Genus	Distinguishing feature	Example	Host	Diseases/pathologies
Alpha-retrovirus	genome <8kb; assembly at cell membrane	avian leukosis virus	birds	malignancies
Beta-retrovirus	intracytoplasmic assembly (B- or D-type)	mouse mammary tumor virus	mice	mammary and ovarian carcinoma; lymphomas
Gamma-retrovirus	genome >8kb; assembly at cell membrane	murine leukemia virus	mice	malignancies
Delta-retrovirus	genomes <9kb; C-type	bovine leukemia virus	cows	malignancies
Epsilon-retrovirus	assembly at cell membrane; hosts: fish	walleye dermal sarcoma virus	fish	solid tumors
Lentivirus	genome >9kb; bar-shaped concentric core	human immunodeficiency virus	humans	immunodeficiency and neurologic disease
Spumavirus	assembly as intracyto-plasmic particles	chimpanzee foamy spumavirus	simians	none apparent

of retroviruses is single-stranded and possesses "positive" polarity similar to that found in messenger RNA (mRNA). Virions (virus particles) contain two 5′ ("five prime"), end-linked, identical copies of the genome RNA, and are therefore said to be diploid.

Table 1. Viruses of the family *Retroviridae*.

Three genes are universally present in the genomes of retroviruses that are capable of replication, such as murine (mouse) leukemia virus. The *gag* (group antigen) gene encodes proteins that make up the nucleocapsid of the virus as well as a matrix layer, the two of which surround the RNA. The *pol* gene (a type of polymerase) encodes reverse transcriptase, which copies the RNA into DNA, and integrase, which integrates the DNA into the host chromosome. Depending on the species, *pol* can also encode protease, a protein that cleaves the initial multiprotein products of retrovirus translation to make functional proteins. Some retroviruses have incorporated viral oncogene sequences. An example of this is reticuloendotheliosis virus strain T. The genome of complex retroviruses, such as HTLV, can contain several other genes that regulate genome expression or replication and are not present in simple retroviruses.

Reverse Transcriptase

Retroviruses follow the same general steps in their replication cycles that are common to other viruses. The steps that differ from other viruses involve the retroviral reverse transcriptase, an enzyme discovered simultaneously by Howard Temin and David Baltimore in 1970. (Temin and Baltimore were awarded the Nobel Prize for this work in 1975.) Reverse transcriptase converts the single-stranded, positive-polarity RNA genome of retrovirus into double-stranded DNA, thereby reversing the typical flow of genetic information (which is from DNA to mRNA). The DNA copy is transported into the nucleus of the host cell, circularized, and integrated into the host chromosome.

This DNA copy of the retrovirus genome is referred to as the provirus or proviral DNA. The genomes of most vertebrates contain abundant numbers of incomplete and complete proviruses (endogenous retroviruses) that

The genomes of simple retroviruses, such as murine leukemia virus (MLV), contain only three genes, *gag, pol,* and *env.* The genomes of complex retroviruses, such as human T-cell lymphotropic virus (HTLV) contain additional genes (e.g., *tax* and *rex)* for regulatory proteins.

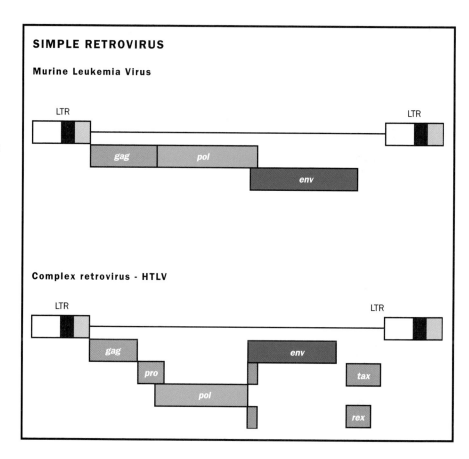

appear to represent remnants of past retroviral infections in germline cells. Proviruses contain structures called long terminal repeats (LTR) at each end. The LTRs contain promoter elements and transcriptional start sites that enable the retroviral genes to be expressed. They can also affect the expression of nearby cellular genes.

Retrovirus Replication Cycle

There are seven steps in the replication cycle of the retrovirus. The first step is attachment, in which the retrovirus uses one of its glycoproteins to bind to one or more specific cell-surface receptors on the host cell. Some retroviruses also employ a secondary receptor, referred to as the co-receptor. Some retroviral receptors and coreceptors have been identified. For example, CD4 and various members of the chemokine receptor family on human T cells (a type of white blood cell) serve as the HIV receptors and coreceptors.

The second and third steps are penetration and uncoating, respectively. Retroviruses penetrate the host cell by direct fusion of the virion envelope with the plasma membrane of the host. Continuation of this fusion process results in the release of the viral capsid directly into the host cell's cytoplasm, where it is partially disrupted.

Step four is replication, which occurs after the retrovirus has undergone partial uncoating. At this stage, the RNA genome is converted by reverse transcriptase into double-stranded DNA. Reverse transcriptase has three enzymatic activities: RNA-directed DNA polymerase makes one

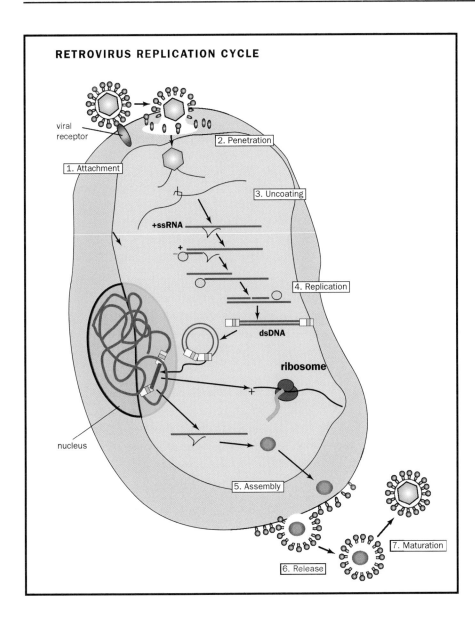

RETROVIRUS REPLICATION CYCLE

viral receptor

1. Attachment

2. Penetration

3. Uncoating

+ssRNA

4. Replication

dsDNA

ribosome

nucleus

5. Assembly

6. Release

7. Maturation

Retrovirus attaches to a host cell membrane, penetrates, and uncoats its genomic RNA. Single-stranded RNA (ssRNA) is used as a template for creation of double-stranded DNA (dsDNA), which is then integrated into the host genome. Once transcribed back into RNA, it codes for viral proteins, is assembled into viral particles, and is released from the cell.

DNA strand, DNA-directed DNA polymerase makes the complementary strand, and RNAse H degrades the viral RNA strand. Reverse transcription is primed by a cellular transfer RNA (tRNA) that is packaged into retrovirus virions. It concludes with the synthesis of a double-stranded copy of the retroviral genome that is termed the "provirus," or proviral DNA.

This proviral DNA is circularized and transported to the host cell's nucleus, where it is integrated, apparently at random, into the genome by means of the retroviral enzyme called integrase. Following integration, the provirus behaves like a set of cellular genes, while the LTRs function as **promoters** that begin transcription back into mRNA. This transcription is carried out by RNA polymerases in the host cell. Transcription of the proviral DNA is also the means of generating progeny RNA. Viral proteins are made in the cytoplasm of the host cell by cellular ribosomes.

The next step (step five) is termed "assembly," in which retrovirus capsids are assembled in an immature form at various locations in the host cell.

promoters DNA sequences to which RNA polymerase binds to begin transcription

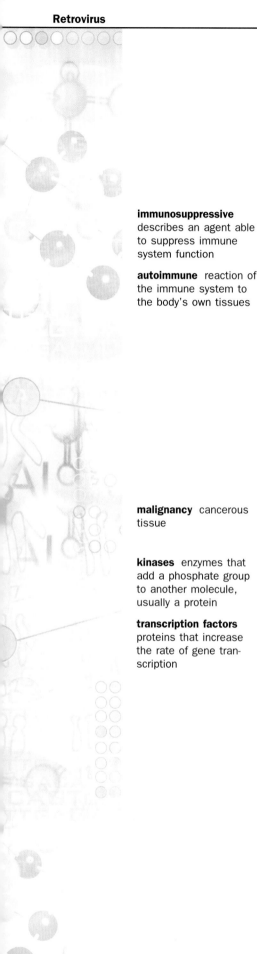

This is followed by an "egress" stage, in which the envelope proteins of retroviruses are acquired by budding from the plasma membrane (cell surface) of the host. Finally, step seven is "maturation." In this step, the Gag and Pol proteins of the retrovirus are cleaved by the retroviral protease, thus forming the mature and infectious form of the virus.

Consequences of Retroviral Infection

Retroviral infection can result in several different outcomes for the virus and the cell. Retroviruses are capable of inducing **immunosuppressive**, **autoimmune**, and neurological illnesses. Some retroviruses, such as the lentiviruses and the spumaviruses, are capable of directly killing cells. Cytopathic (cell-killing) effects in infected T cells and cells in the brain may account for the profound immune deficiencies and neurological diseases induced by HIV and related lentiviruses.

Retroviruses are also capable of inducing latent infections, in which the virus is dormant, or persistent infections, in which low levels of the virus are continuously produced. These capabilities explain the life-long nature of retroviral infections, and render the diseases induced by these pathogens extremely difficult to treat.

Retroviruses and Cancer

Retroviruses are among several types of viruses that can induce cancer in the host organism. So-called slowly transforming viruses are exemplified by human T-lymphotropic virus (HTLV), which causes leukemia (a type of blood cancer) in humans. These viruses induce **malignancy** by a process called insertional mutagenesis. The initial event is thought to be retroviral integration near, and subsequent activation of, a cellular oncogene (*c-onc*). Examples of *c-onc* include genes for growth factors, protein **kinases**, and **transcription factors**. Harold Varmus and Michael Bishop won the Nobel Prize for physiology or medicine in 1989 for their contributions to the discovery of oncogenes.

When a malignancy is triggered, tumors appear only after a long latent period of months or years, and these tumors are typically clonal in origin. That is, they arise by the rare transformation of a single cell. HTLV-1 is highly prevalent in people living in Japan, the Caribbean, and Africa, areas where approximately one percent of adults are infected. About one to three percent of infected individuals will eventually develop adult T-cell leukemia after an incubation period, which is usually several decades long. HTLV stimulates T-cell proliferation that could favor mutational events leading to cell transformation.

Acutely transforming retroviruses contain a viral oncogene (*v-onc*) and induce polyclonal cancers (that is, many different cancer cells are derived in multiple transforming events) at high efficiency within a short time frame (weeks). The *v-onc* are derived by incorporation and modification (that is, by deletion of introns, mutations, and other such processes) of host-cell oncogenes. The *v-onc* are often expressed in great quantity, due to the highly active viral LTRs. Most acutely transforming retroviruses are replication-defective, because incorporation of the oncogene deletes an essential gene or genes. They therefore require a helper virus to prop-

immunosuppressive describes an agent able to suppress immune system function

autoimmune reaction of the immune system to the body's own tissues

malignancy cancerous tissue

kinases enzymes that add a phosphate group to another molecule, usually a protein

transcription factors proteins that increase the rate of gene transcription

agate. An exception is Rous sarcoma virus, whose genome retains enough of the structural gene sequences to remain capable of replication. SEE ALSO DNA LIBRARIES; EVOLUTION OF GENES; GENE THERAPY; HIV; ONCOGENES; OVERLAPPING GENES; REVERSE TRANSCRIPTASE; RNA; RNA POLYMERASES; VIRUS.

Robert Garry

Bibliography

Varmus, Harold E. "Form and Function of Retroviral Proviruses." *Science* 216, no. 4548 (1982): 812–820.

Weinberg, Robert A. "How Cancer Arises." *Scientific American* 275, no. 3 (1996): 62–70.

Reverse Transcriptase

Reverse transcriptase is the replication enzyme of retroviruses. Because it **polymerizes** DNA precursors, reverse transcriptase is a DNA polymerase. However, whereas cellular DNA polymerases use DNA as a template for making new DNAs, reverse transcriptase uses the single-stranded RNA in retroviruses as the template for synthesizing viral DNA. This unusual process of making DNA from RNA is called "reverse transcription" because it reverses the flow of genetic information (from DNA to RNA, rather than from RNA to DNA found in transcription). Because reverse transcriptase is essential for retroviruses such as HIV-1 (the virus that causes AIDS), it is the target of many antiretroviral therapeutics. Reverse transcriptase is also a molecular tool used in the cloning of genes and the analysis of gene expression.

polymerizes links together similar parts to form a polymer

Discovery

Retroviruses were originally known as RNA tumor viruses because they have RNA, not DNA, genomes, and because they were the first viruses recognized to cause certain cancers in animals. At the middle of the twentieth century, Howard Temin was interested in understanding how RNA tumor viruses cause cancer. One finding that interested him was the genetic-like stability of the uncontrolled cell growth caused by these viruses. It was known then that certain bacterial viruses, called phages, could integrate their DNA into their hosts' chromosomes and persist as stable genetic elements known as prophages. By analogy, Temin proposed the provirus hypothesis, which suggests that RNA tumor viruses can cause permanent alterations to cells by integrating into host chromosomes. In order for this to occur, Temin suggested that **virion** RNAs were first converted into DNAs, which could then become integrated.

virion virus particle

The chemistry of using RNA as a **template** for DNA seemed possible. However, reverse transcription was at odds with the then-popular central dogma of molecular biology, advanced by Francis Crick, which maintained that genetic information flowed unidirectionally from DNA to RNA to protein. RNA tumor viruses were RNA viruses, so it was assumed that their replication involved RNA polymerases, as had been demonstrated for other RNA viruses, and not a DNA polymerase. Because his proposal of a reverse

template a master copy

Reverse transcriptase uses a single-stranded RNA template to create a double-stranded DNA.

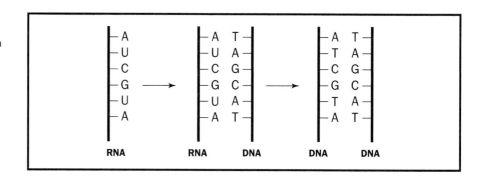

flow of genetic information from RNA to DNA seemed heretical, and because the experimental techniques needed to test this idea were not yet developed, Temin and his hypothesis were rebuffed for many years.

The biochemical proof for reverse transcription finally arrived in 1970 when two separate research teams, one led by Temin and the other by David Baltimore, simultaneously discovered the elusive RNA-copying DNA polymerase in purified virions. In 1975 Temin and Baltimore shared the Nobel Prize in physiology or medicine for their discovery of reverse transcriptase.

Laboratory Uses of Reverse Transcriptase

Reverse transcriptase went on to play a critical role in the molecular revolution of the late 1970s and 1980s, especially in the fields of gene discovery and biotechnology. Genes can often be discovered most easily by isolating and analyzing the messenger RNA (mRNA) production in a cell. Reverse transcriptase allowed the synthesis of cDNA, or complementary copies of messenger RNAs. The cDNA can then be expressed in a model organism such as *Escherichia coli*, and the protein it codes for can then be made in abundance. The cloning of cDNA was instrumental to gene discovery in the later part of the twentieth century. Using cDNA copies of genes is necessary when bacteria are used to produce human protein-based pharmaceuticals. This is because bacteria lack the machinery necessary to recognize unspliced genes, but bacteria can use cDNAs to direct the synthesis of human or other higher organism proteins.

mutations changes in DNA sequences

Even though the human genome sequence was reported in 2001, copying RNAs with reverse transcriptase remains important. One reason for this is that some human diseases result from **mutations** in genes whose products act to adjust the sequences of RNAs after transcription but before protein synthesis. Thus, even though prototype human sequences are available, it appears likely that molecular diagnostics will include screening cDNA copies of individual people's RNAs. Other uses of cDNA include generating probes to screen microarrays to assess variation in gene expression and regulation.

Reverse Transcriptase and AIDS

Soon after AIDS was recognized in the early 1980s, Luc Montagneer of France and, subsequently, the American Robert Gallo determined that the causative agent was a retrovirus. Like other retroviruses, HIV-1 contains reverse transcriptase and must generate DNA. Differences between reverse

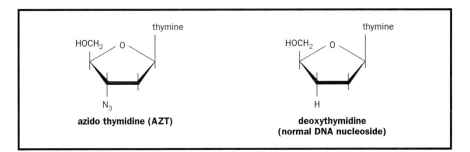

azido thymidine (AZT)

deoxythymidine
(normal DNA nucleoside)

AZT becomes
incorporated into the DNA
chain formed by reverse
transcriptase, but the N_3
groups block further
replication.

transcriptase and cellular DNA polymerases in the sorts of DNA precursors (**nucleosides**) that they can utilize have been exploited to develop drugs that are selectively toxic to HIV-1.

nucleosides building blocks of DNA or RNA, composed of a base and a sugar

Azidothymidine (AZT) is an example of a nucleoside analog DNA precursor that can serve as a reverse transcriptase "suicide inhibitor," because AZT incorporation into viral DNA prevents later steps in viral replication. However, the effectiveness of these sorts of drugs is limited by several factors. AZT is occasionally incorporated into cellular DNA, which contributes to the toxicity some patients experience when treated with reverse transcriptase inhibitors. Additionally, reverse transcriptase inhibitor resistance often develops during antiretroviral therapy. This resistance results from reverse transcriptase's high error rate, which generates a remarkable amount of genetic variation within HIV populations. If some viral genetic variants are less sensitive to antivirals than other variants, the resistant mutants will replicate during antiviral therapy. Despite these complications, reverse transcriptase inhibitors remain important components of the combined antiviral regimen that has dramatically lengthened the lives of many HIV-infected patients since the mid-1990s.

Reverse Transcription and the Human Genome

When reverse transcriptase was first described, it was believed to be a peculiarity of retroviruses. However, researchers now know that reverse transcription also occurs during the replication of the DNA virus hepatitis B, and that RNA-copying DNA polymerases function within human cells. One of these host reverse transcriptases is telomerase, an enzyme that helps maintain chromosome ends.

Other human reverse transcriptases are parts of **endogenous** retroviruses and retroelements, such as those that encoded the majority of the repetitive "junk" DNA in human chromosomes. Many of these retroelements integrated their DNAs into our chromosomes so long ago that they predate human speciation. Because of this, molecular **phylogeneticists** can use sites of retroelement insertions to determine the lineages and ancestral relationships of species. Thus, while retroviruses, in the form of HIV-1, represent one of the newest diseases of humans, the prevalence of other retrovirus-like elements in our genomes demonstrates the long-standing relationship of humans with reverse transcribing elements. SEE ALSO DNA MICROARRAYS; DNA POLYMERASES; EVOLUTION OF GENES; HIV; NUCLEOTIDE; RETROVIRUS; TELOMERE; TRANSCRIPTION; TRANSPOSABLE GENETIC ELEMENTS.

endogenous derived from inside the organism

phylogeneticists scientists who study the evolutionary development of a species

Alice Telesnitsky

41

Bibliography

Kazazian, Haig H., Jr. "L1 Retrotransposons Shape the Mammalian Genome." *Science* 289, no. 5482 (2000): 1152–1153.

Varmus, H. "Reverse Transcription." *Scientific American* 257, no. 3 (1987): 56–59.

Ribosome

Ribosomes are the cellular **organelles** that carry out protein synthesis, through a process called **translation**. They are found in both **prokaryotes** and **eukaryotes**, these molecular machines are responsible for accurately translating the linear genetic code, via the messenger RNA, into a linear sequence of amino acids to produce a protein. All cells contain ribosomes because growth requires the continued synthesis of new proteins. Ribosomes can exist in great numbers, ranging from thousands in a bacterial cell to hundreds of thousands in some human cells and hundreds of millions in a frog ovum. Ribosomes are also found in **mitochondria** and **chloroplasts**.

Structure

The ribosome is a large ribonucleoprotein (RNA-protein) complex, roughly 20 to 30 nanometers in diameter. It is formed from two unequally sized subunits, referred to as the small subunit and the large subunit. The two subunits of the ribosome must join together to become active in protein synthesis. However, they have distinguishable functions. The small subunit is involved in decoding the genetic information, while the large subunit has the catalytic activity responsible for peptide bond formation (that is, the joining of new amino acids to the growing protein chain).

In prokaryotes, the small subunit contains one RNA molecule and about twenty different proteins, while the large subunit contains two different RNAs and about thirty different proteins. Eukaryotic ribosomes are even more complex: the small subunit contains one RNA and over thirty proteins, while the large subunit is formed from three RNAs and about fifty proteins. Mitochondrial and chloroplast ribosomes are similar to prokaryotic ribosomes.

In spite of its complex composition, the architecture of the ribosome is very precise. Even more remarkable, ribosomes from all organisms, ranging from bacteria to humans, are very similar in their form and function. Recent breakthroughs in studies of ribosome structure, using techniques such as scanning, cryo-electron microscopy, and X-ray crystallography, have provided scientists with highly refined structures of this complex organelle. One particularly exciting conclusion from studies of the large subunit is that it is ribosomal RNA (rRNA), and not protein, that provides the **catalytic** activity for peptide bond formation. That is, it forms the chemical linkage between the amino acids of the growing protein molecule.

Synthesis

The synthesis of ribosomes is itself a very complex process, requiring the coordinated output from dozens of genes encoding ribosomal proteins and rRNAs. Ribosomes are assembled from their many component parts in an orderly pathway. In eukaryotes, rRNA synthesis and most of the assembly

organelles membrane-bound cell compartments

translation synthesis of protein using mRNA code

prokaryotes single-celled organisms without a nucleus

eukaryotes organisms with cells possessing a nucleus

mitochondria energy-producing cell organelle

chloroplasts the photosynthetic organelles of plants and algae

catalytic describes a substance that speeds a reaction without being consumed

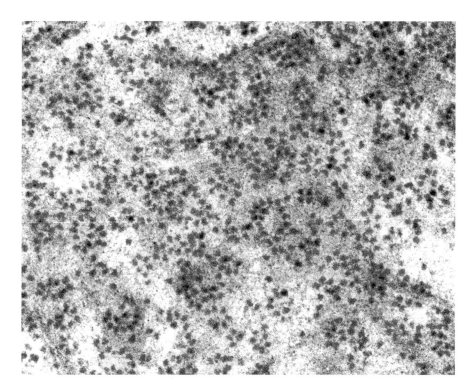

Ribosomes from a liver cell are represented by the darkened ares of the magnified image. These organelles contain RNA and use it for protein synthesis.

steps occur in a structure within the nucleus called the nucleolus. Eukaryotic ribosome synthesis is especially complicated, because the ribosomal proteins themselves are made by ribosomes in the cytoplasm (that is, outside of the nucleus), so they then must be imported into the nucleolus for assembly onto the nucleolus-derived rRNA. Once assembled, the nearly complete ribosomal subunits are then exported out of the nucleus and back into the cytoplasm for the final steps of assembly.

The exact details of the **in vivo** ribosome assembly pathway (the process of ribosome assembly within the living cell) are still under investigation. Assembly in eukaryotic cells involves not only the components of the mature particles, but also dozens of auxiliary factors that promote the efficient and accurate construction of the ribosome during its assembly. However, bacterial ribosomes can be constructed **in vitro** using purified ribosomal proteins and rRNAs. These ribosomes appear to function normally in in vitro translation reactions.

in vivo "in life"; in a living organism, rather than in a laboratory apparatus

in vitro "in glass"; in lab apparatus, rather than within a living organism

Ribosome Function

Translation of messenger RNA (mRNA) by ribosomes occurs in the cytoplasm. In bacterial cells, ribosomes are scattered throughout the cytoplasm. In eukaryotic cells, they can be found both as free ribosomes and as bound ribosomes, their location depending on the function of the cell. Free ribosomes are found in the cytosol, which is the fluid portion of the cytoplasm, and are responsible for manufacturing proteins that will function as soluble proteins within the **cytoplasm** or form structural elements, including the cytoskeleton, that are found within the cytosol.

cytoplasm the material in a cell, excluding the nucleus

Bound ribosomes are attached to the outside of a membranous network called the **endoplasmic reticulum** to form what is termed the "rough" endoplasmic reticulum. Proteins made by bound ribosomes are intended to be

endoplasmic reticulum network of membranes within the cell

incorporated into membranes, or packaged for storage, or exported outside of the cell. Ribosomes exist either as a single ribosome (that is, one ribosome translating an mRNA) or as polysomes (two or more ribosomes sequentially translating the same mRNA in order to make multiple copies of the same protein).

Ribosomes have the critical role of mediating the transfer of genetic information from DNA to protein. Ribosomes translate this code using an intermediary, the messenger RNA, which is a copy of the DNA that can be interpreted by ribosomes. To begin translation, the small subunit first identifies, with the help of other protein factors, the precise point in the RNA sequence where it should begin linking amino acids, the building blocks of protein. The small subunit, once bound to the mRNA, is then joined by the large subunit and translation begins. The amino acid chain continues to grow until the ribosome reaches a signal that instructs it to stop.

Many of the antibiotics used in humans and other animals to treat bacterial infections specifically inhibit ribosome activity in the disease-causing bacteria, without affecting ribosome function in the host-animal's cells. These antibiotics work by binding to a protein or RNA target in the bacterial ribosome and inhibiting translation. In recent years, the misuse of antibiotics has resulted in the natural selection of bacteria that are resistant to many of these antibiotics, either because they have mutations in the antibiotic's target in the ribosome or because they have acquired a mechanism for excluding or inactivating the antibiotic. SEE ALSO CELL, EUKARYOTIC; RIBOZYME; RNA; TRANSLATION.

Janice Zengel

Bibliography

Frank, Joachim. "How the Ribosome Works." *American Scientist* 86 (1998): 428–439

Garrett, Robert A., et al, eds. *The Ribosome: Structure, Function, Antibiotics, and Cellular Interactions.* Washington, DC: ASM Press, 2000

Karp, Gerald. *Cell and Molecular Biology: Concepts and Experiments*, 3rd ed. New York: John Wiley & Sons, 2002.

Ribozyme

Ribozymes are RNA molecules that catalyze chemical reactions. Most biological processes do not happen spontaneously. For example, the cleavage of a molecule into two parts or the linkage of two molecules into one larger molecule requires **catalysts**, that is, helper molecules that make a reaction go faster. The majority of biological catalysts are proteins called enzymes. For many years scientists assumed that proteins alone had the structural complexity needed to serve as specific catalysts in cells, but around 1980 the research groups of Tom Cech and Sidney Altman independently discovered that some biological catalysts are made of RNA. These two scientists were honored with the Nobel Prize in chemistry in 1989 for their discovery.

Structure and Function

The RNA catalysts called ribozymes are found in the nucleus, mitochondria, and chloroplasts of **eukaryotic** organisms. Some viruses, including

catalysts substances that speed up a reaction without being consumed (e.g., enzyme)

eukaryotic describing an organism that has cells containing nuclei

several bacterial viruses, also have ribozymes. The ribozymes discovered to date can be grouped into different chemical types, but in all cases the RNA is associated with metal ions, such as magnesium (Mg2+) or potassium (K+), that play important roles during the catalysis. Almost all ribozymes are involved in processing RNA. They act either as molecular scissors to cleave precursor RNA chains (the chains that form the basis of a new RNA chain) or as "molecular staplers" that **ligate** two RNA molecules together. Although most ribozyme targets are RNA, there is now very strong evidence that the linkage of amino acids into proteins, which occurs at the ribosome during **translation**, is also catalyzed by RNA. Thus, the ribosomal RNA is itself also a ribozyme.

In some ribozyme-catalyzed reactions, the RNA cleavage and ligation processes are linked. In this case, an RNA chain is cleaved in two places and the middle piece (called the intron) is discarded, while the two flanking RNA pieces (called exons) are ligated together. This reaction is called splicing. Besides ribozyme-mediated splicing, which involves RNA alone, there are some splicing reactions that involve RNA-protein complexes. These complexes are called small nucleus ribonucleoprotein particles, abbreviated as snRNPs. This class of splicing is a very common feature of messenger RNA (mRNA) processing in "higher" eukaryotes such as humans. It is not yet known if snRNP-mediated splicing is catalyzed by the RNA components. Note also that some RNA splicing reactions are catalyzed by enzymes made of only protein.

Some precursor RNA molecules have a ribozyme built into their own intron, and this ribozyme is responsible for removal of the intron in which it is found. These are called self-splicing RNAs. After the splicing reaction is complete, the intron, including the ribozyme, is degraded. In these cases, each ribozyme works only once, unlike protein enzymes that catalyze a reaction repeatedly. Examples of self-spliced RNAs include the ribosomal RNAs of **ciliated protozoa** and certain mRNAs of yeast mitochondria.

Some RNA viruses, such as the hepatitis delta virus, also include a ribozyme as part of their inherited RNA molecule. During replication of the viral RNA, long strands containing repeats of the RNA **genome** (viral genetic information) are synthesized. The ribozyme then cleaves the long **multimeric** molecules into pieces that contain one genome copy, and fits that RNA piece into a virus particle.

Other ribozymes work on other RNA molecules. One ribozyme of this type is RNase P, which consists of one RNA chain and one or more proteins (depending on the organism). The catalytic mechanism of RNase P has been especially well-studied in bacteria. This ribozyme processes precursor transfer RNA (tRNA) by removing an extension from the 5-prime end, to create the 5-prime end of the "mature" tRNA (the two ends of an RNA molecule are chemically distinct and are called the 5-prime and 3-prime ends, referring to specific carbons in the sugar moiety of the terminal nucleotides). When the RNA molecule from bacterial RNase P is purified away from its protein, it can still cleave its precursor tRNA target, albeit at a very slow rate, proving that the RNA is the catalyst. Nevertheless, the protein(s) in RNase P also has important functions, such as maintenance of the proper conformation of the RNase P RNA and interaction with the precursor tRNA.

ligate join together

translation synthesis of protein using mRNA code

ciliated protozoa single-celled organism possessing cilia, short hair-like extensions of the cell membrane

genome the total genetic material in a cell or organism

multimeric describes a structure composed of many similar parts

Relics of an "RNA World"

Many biologists hypothesize that ribozymes are vestiges of an ancient, prebiotic world that predated the evolution of proteins. In this "RNA world," RNAs were the catalysts of such functions as replication, cleavage, and ligation of RNA molecules. Proteins are hypothesized to have evolved later, and as they evolved they took over functions previously performed by RNA molecules. This may have happened because proteins are more versatile and efficient in their catalytic functions.

In today's world, most processing of precursor tRNA is performed by the ribozyme RNase P, as described above, but in some chloroplasts, this function is performed by a protein that apparently contains no RNA. This may be an example of the evolution of protein enzymes that replace ribozymes.

Intensive studies of ribozymes have provided rules for how they recognize their targets. Based on these rules, it has been possible to alter ribozymes to recognize and cleave new targets in RNA molecules that are normally not subject to ribozyme cleavage. These results raise the exciting possibility of using ribozymes for human therapy. For example, the abundance of disease-causing RNA molecules such as HIV, the cause of AIDS, could be reduced with artificial ribozymes. Considerable success has been achieved in testing these ribozymes in model cells. However, the biggest question remaining to be solved is how these potential "disease-fighting" ribozymes can be introduced into a patient and taken up by the appropriate cells. SEE ALSO Evolution, Molecular; Proteins; RNA; RNA Processing.

Lasse Lindahl

Bibliography

Cech, T. R. "RNA as an Enzyme." *Scientific American* 255 (1986): 64–75.

Karp, Gerald. *Cell and Molecular Biology*, 3rd ed. New York: John Wiley & Sons, 2002.

RNA

Ribonucleic acid (RNA) molecules, which are linear chains (or polymers) of ribonucleotides, perform a number of critical functions. Many of these functions are related to protein synthesis. Some RNA molecules bring genetic information from a cell's chromosomes to its **ribosomes**, where proteins are assembled. Others help ribosomes translate genetic information to assemble specific sequences of amino acids.

ribosomes protein-RNA complexes at which protein synthesis occurs

Molecular Structure

Ribonucleotides, the building blocks of RNA, are molecules that consist of a nitrogen-containing base, a phosphate group, and ribose, a five-carbon sugar. The nitrogen-containing base may be adenine, cytosine, guanine, or uracil. These four bases are abbreviated as A, C, G, and U.

RNA is similar to deoxyribonucleic acid (DNA), another class of nucleic acid. However, DNA nucleotides contain deoxyribose, not ribose, and they use the nitrogen-containing base thymine (T), not uracil, along with adenine, cytosine, and guanine.

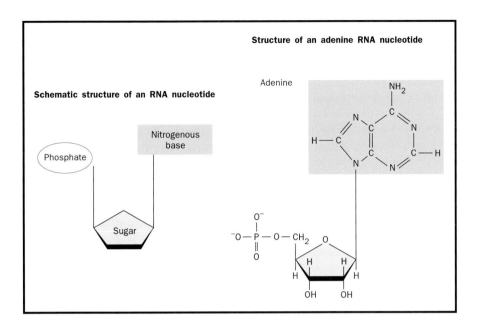

Structure of an adenine RNA nucleotide

Schematic structure of an RNA nucleotide

Adenine

Nitrogenous base

Phosphate

Sugar

Nucleotides link to form RNA chains. Adenine is one of the four bases found in RNA. Adapted from Robinson, 2001.

The nucleotides in DNA and RNA molecules are linked together to form chains. The link between two nucleotides is between a phosphate group attached to the fifth (5′ or "five prime") carbon of the sugar on one nucleotide and a hydroxyl group on the third (3′ or "three prime") carbon of the sugar on the other. The link is called a 5′–3′ phosphodiester bond.

RNA, therefore, can be described as a chain of ribose sugars linked together by phosphodiester bonds, with a base protruding from each sugar, as shown in the figure below. The 5′–3′ linkage gives RNA directionality, or polarity, and results in its having two ends with different chemical structures. The 5′ end usually has one or three free phosphate groups, and the 3′ end usually has a free hydroxyl group.

Whereas DNA is usually double-stranded, with the bases on one strand pairing up with those on the other, RNA usually exists as single chains of nucleotides. The bases in RNA do, however, follow Watson-Crick base-pair rules: A and U can pair with each other, as can G and C. There is usually extensive pairing of bases within a single strand of RNA.

RNA strands fold, with the bases in one part of the strand pairing with the bases in another. Folding can create both "secondary" and "tertiary" structures. Secondary structures are those that can be described in two dimensions and that can be thought of as simple loops or helices. Tertiary structures are complex, three-dimensional shapes.

The most common secondary structures, "hairpins," "loops," and "pseudo-knots," are shown in the figure below. Such secondary structures are formed when **hydrogen bonds** form between bases in the nucleotides and by the stacking of bases to form helical structures.

Tertiary structures usually involve interactions between nucleotides that are distant from each other along an RNA strand. Such interactions may arise from hydrogen bonding between bases, as in regular Watson-Crick base pairing, or from interactions among other chemical groups in the nucleotides. Some RNA molecules, such as ribosomal RNA (rRNA)

hydrogen bonds weak bonds between the H of one molecule or group and a nitrogen or oxygen of another

The molecular structures of the four RNA bases. Adapted from Robinson, 2001.

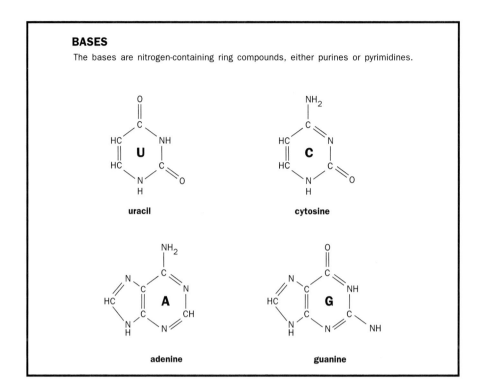

BASES

The bases are nitrogen-containing ring compounds, either purines or pyrimidines.

uracil

cytosine

adenine

guanine

and transfer RNA (tRNA), have structures that are very complex. In structure they resemble proteins more than they do DNA.

To understand the function of a given RNA molecule, scientists often need to know its structure. There are three general strategies for analyzing RNA structure. First, using the relatively simple base-pairing rules for RNA and the basic principles of **thermodynamics**, computers can be used to predict secondary RNA structure, although not always with complete success.

Second, researchers can analyze RNA molecules from various organisms and compare those molecules that have the same function. Even when the nucleotide sequences vary between species, important structures are usually preserved.

Third, the structure of an RNA molecule can be determined experimentally, using **enzymes** to cut it or chemicals to modify it. Some enzymes and chemicals cut or modify only nonpaired, single-stranded portions of the RNA molecule, allowing researchers to identify double-stranded regions by examining which ones remain uncut and unmodified.

Despite the usefulness of each of these methods, none can provide a complete and accurate three-dimensional structure. A more complete determination of structure can be achieved by the biophysical methods of X-ray crystallography and nuclear magnetic resonance.

Synthesis

RNA molecules are synthesized by enzymes known as RNA polymerases in a process called transcription. Usually, one strand of a double-stranded DNA molecule is used as a **template** for the RNA. The order of ribonucleotides that are assembled to form the RNA molecule is determined by the order of the deoxyribonucleotides in the DNA strand. The genetic information

thermodynamics process of energy transfers during reactions, or the study of these processes

enzymes proteins that control a reaction in a cell

template a master copy

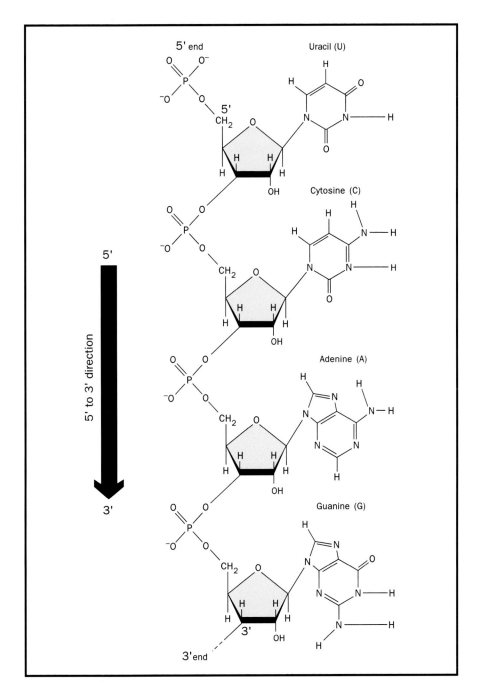

RNA nucleotides link together to form a chain. The phosphate group, attached to the 5′ carbon of ribose, links to the OH group at the 3′ position. This chain itself has a 5′ end and a 3′ end, as indicated.

in the DNA sequence is thus reproduced in the RNA molecule. Sometimes, but rarely, an RNA molecule is synthesized using another RNA molecule as the template.

Often, when RNA molecules are synthesized, they are in a form that prevents them from carrying out their function. To become functional, they must undergo processing, which can involve removing segments of the strands or modifying specific nucleotides. The link between a base and a ribose may be altered, or extra chemical groups may be added to the bases or ribose molecules. Many RNA molecules are associated with proteins during or after their synthesis. Together, the RNA and protein are referred to as RNA-protein particles (RNPs).

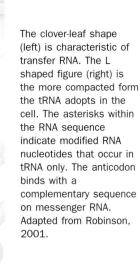

The clover-leaf shape (left) is characteristic of transfer RNA. The L shaped figure (right) is the more compacted form the tRNA adopts in the cell. The asterisks within the RNA sequence indicate modified RNA nucleotides that occur in tRNA only. The anticodon binds with a complementary sequence on messenger RNA. Adapted from Robinson, 2001.

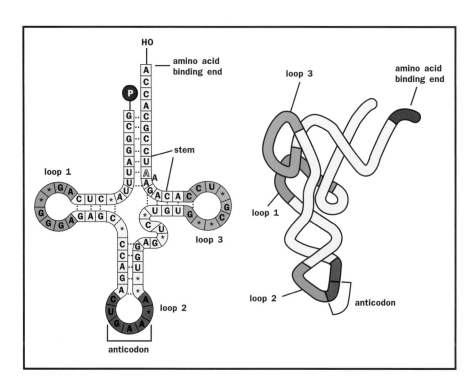

eukaryotes organisms with cells possessing a nucleus

In **eukaryotes**, RNA that is encoded by nuclear chromosomes is synthesized in the nucleus. The processing and assembly of many small RNA molecules in higher eukaryotes is accomplished in Cajal bodies, which are coiled structures in the nucleus that were identified more than 100 years ago but that have begun to be investigated in detail only recently. The synthesis of those RNA molecules that are components of ribosomes occurs in the nucleolus, a part of the nucleus. RNA synthesis and processing also occurs in the mitochondria and chloroplasts, when the RNA will be used in those organelles.

cytoplasm the material in a cell, excluding the nucleus

prokaryotes single-celled organisms without a nucleus

After being processed and assembled, RNPs either remain in the nucleus or are exported to the **cytoplasm** through the nuclear pores. Some are also exported and modified in the cytoplasm and then imported back into the nucleus. In **prokaryotes**, where there is no nucleus, the synthesis and processing of RNA, as well as the assembly of RNPs, occurs in cytoplasm.

Function

Almost all types of RNA play a role in translation, which is the process of protein synthesis. Translation requires three types of RNA: messenger RNA (mRNA), which ranges in length from a few hundred to many thousands of nucleotides; tRNA, which is 75 to 85 nucleotides long; and rRNA, which is 1,500 to 4,000 nucleotides long.

Molecules of mRNA, each of which contains a copy of at least one gene, are the intermediates between DNA and protein. These mRNA molecules bring the genetic code from the DNA, which is in the nucleus, to ribosomes, which are in the cytoplasm. They attach to the ribosomes and determine the order in which amino acids are assembled to synthesize a protein. Of the three types of RNA required for translation, mRNA molecules have the simplest structure.

Next, tRNA molecules function as adapters that help translate the nucleotide sequences in mRNA into amino acid sequences, so specific proteins can be constructed. There are many different types of tRNA, each of which is capable of binding to one of the twenty amino acids that are the building blocks of proteins.

Finally, rRNA molecules, which account for most of a ribosome's mass, are, according to recent experiments, the part of the ribosome responsible for linking amino acids into a growing protein chain. Ribosomes, the organelles that assemble a particular sequence of amino acids to form proteins, contain three or four different molecules of rRNA, along with at least fifty different proteins.

Both rRNA and tRNA are stable forms of RNA that last through several cell divisions. In contrast, mRNA is normally unstable, with a lifetime that can be as short as a few minutes. This instability has probably evolved because it lets cells quickly stop synthesizing proteins that are no longer needed. In some cases, enzymes called ribonucleases (RNases) actively degrade certain mRNA molecules. For example, mRNA that encodes a particular protein regulating the **cell cycle** is degraded when the protein has carried out its function.

cell cycle sequence of growth, replication and division that produces new cells

In certain cells, mRNA can exist in a stable form for decades. When egg cells are formed, for example, some of the mRNA in the cells is associated with "storage proteins" and lasts until after the eggs are fertilized. During embryonic development, this maternal mRNA becomes activated for translation and associates with translating ribosomes. It usually decays after it has been used to produce a certain amount of protein.

Less Common Types of RNA

Several types of less abundant, small RNA molecules perform essential functions in both the nucleus and the cytoplasm. All organisms contain cytoplasmic RNPs that are involved in exporting proteins from cells. During the synthesis of proteins that are destined to be exported, the ribosome and mRNA associate with an "export-RNP," which helps them dock at an export pore in the cell membrane. As it formed, the protein is threaded through the membrane to the outside of the cell. In eukaryotes, this same strategy is used to transport proteins into the **endoplasmic reticulum**, where some newly synthesized proteins are sorted and modified.

endoplasmic reticulum network of membranes within the cell

RNase P is another RNP found in all forms of life. This RNA-containing enzyme helps turn precursor tRNA into mature tRNA molecules. It does so by cleaving a section off the 5′ end of the precursor molecules.

Small nucleolar RNAs, which are known as snoRNAs and which are found in the nucleoli of eukaryotes and in **Archaea**, are required for the processing of precursor rRNA. During the assembly of new ribosomes, snoRNAs help remove regions of the precursor molecules and modify specific nucleotides.

Archaea one of three domains of life, a type of cell without a nucleus

Often, mRNA molecules in eukaryotes and in Archaea contain sequences that do not code for amino acids. These sequences, called introns, must be

a cell can avoid creating proteins that may be derived from viruses, albeit at the risk of turning off one or more of its own genes at the same time.

Research Uses of RNA Interference

Because of its ability to turn off individual gene expression, RNA interference provides a remarkably precise tool for studying the effects of individual genes. There are several ways to deliver dsRNA to cells. It can be injected into a single cell or placed in a viral chromosome that infects the cells being studied. Roundworms will absorb dsRNA if they are soaked in a solution containing it, or if they eat bacteria that contain it.

RNA interference has several advantages over the alternative way of "knocking out" a gene, called gene targeting. Unlike gene targeting, administration of dsRNA does not require long and laborious breeding of the target organism carrying the knockout. Even more importantly, the dsRNA knockout is temporary and can be induced at any stage of the life cycle, rather than exerting its effect throughout life, as with gene targeting. This allows short-term studies of gene effects, a feature particularly valuable for studying development, for instance.

New Developments in dsRNA

Recent research has also shown that a class of similar dsRNA fragments, called small temporal RNAs, play important roles in development in the roundworm, fruit fly, and other animals. Although little is so far known about them, these fragments are made by dicer from the cell's own RNA as a normal part of the developmental process and appear to help control **gene expression**. This is an exception to the statement that the presence of dsRNA signals a threat to the cell; how these are distinguished from threatening dsRNA is not yet known. SEE ALSO ANTISENSE NUCLEOTIDES; FRUIT FLY: *DROSOPHILA*; NUCLEASES; RNA; POST-TRANSLATIONAL CONTROL; RNA PROCESSING; ROUNDWORM: *CAENORHABDITIS ELEGANS*; TRANSPOSABLE GENETIC ELEMENTS; VIRUS.

Richard Robinson

gene expression use of a gene to create the corresponding protein

Bibliography

Ambros, Victor. "Development: Dicing up RNAs." *Science* 293 (2001): 811–813.

Gura, Trisha. "A Silence that Speaks Volumes." *Nature* 404 (2000): 804–808.

Lin, Rueyling, and Leon Avery. "RNA Interference. Policing Rogue Genes." *Nature* 402 (1999): 128–129.

RNA Polymerases

RNA polymerases are enzyme complexes that synthesize RNA molecules using DNA as a **template**, in the process known as transcription. The RNAs created by transcription are either used as is (as ribosomal RNAs, transfer RNAs, or other types), or serve to guide the **synthesis** of a protein (as messenger RNAs). The word "polymerase" derives from "-ase," a suffix indicating an enzyme, and "polymer," meaning a large molecule composed of many similar parts, in this case the RNA nucleotides A, U, C, and G (abbreviations for adenine, uracil, cytosine, and guanine).

template a master copy

synthesis creation

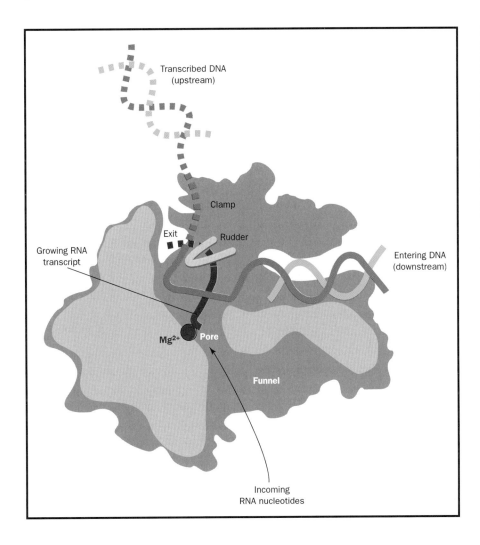

RNA polymerase use one side of the DNA double helix (blue) to assemble RNA nucleotides into an RNA transcript. The internal structure of the polymerase, shown in cut-away view, helps separate the DNA strands. A magnesium ion (Mg^{2+}) helps catalyze the addition of RNA nucleotides into the growing chain. Adapted from <http://www.euchromatin.org/1844-1-med.gif>.

Types of Polymerases

Prokaryotic organisms (Eubacteria and Archaea) have only one type of RNA polymerase. **Eukaryotic** organisms (animals, plants, fungi, and protists) have three types, called pol I, II, and III, and each transcribes a different set of genes. Pol I synthesizes RNA for the large subunit of the ribosome (the protein-making machinery of the cell), and one piece of RNA for the small subunit. Pol II creates messenger RNAs, which provide a template for protein synthesis. Pol II also creates numerous small nuclear RNAs (snRNAs), which modify RNAs after they are formed. Pol III synthesizes transfer RNAs (tRNAs), the RNA for the small subunit of the ribosome, and other snRNAs.

The three eukaryotic polymerases can be distinguished in the laboratory by the degree to which they are inhibited by the alpha-amanitin poison from the mushroom *Amanita phalloides*. Pol I is completely resistant to its effects, pol III is moderately sensitive, and pol II is highly sensitive. (The reason this poison is so deadly is precisely because it inhibits RNA polymerase.)

Each eukaryotic RNA polymerase is composed of a dozen or more subunits. Some of these are small and unique to each type, but the two largest subunits are similar among all three polymerases, and similar as well to the two largest prokaryotic subunits. This is clear evidence that all of them

prokaryotic to come

eukaryotic desribing an organism that has cells containing nuclei

evolved from the same original polymerase. These shared subunits are thought to form the functional core of the polymerases, while the smaller subunits may provide the gene specificity of each type.

Transcription

Transcription begins when RNA polymerase binds to the DNA double helix. This occurs at a site just "upstream" of the gene to be transcribed, called the promoter site. In eukaryotes, RNA polymerase is directed to the promoter site by **transcription factors**, proteins that bind to the DNA and provide a docking site for attachment of the polymerase enzyme. Once RNA polymerase binds to the DNA at the promoter, transcription can begin.

During transcription, the polymerase unwinds a portion of the double-stranded DNA, exposing the DNA template strand that will be copied into RNA. Individual RNA nucleotides enter the enzyme complex, and are paired with the DNA. C pairs with G, T (on DNA) pairs with A, and A (on DNA) pairs with U. Nine DNA-RNA nucleotide pairs exist within the polymerase molecule at any one time. After each new RNA nucleotide is paired, it is linked to the preceding RNA nucleotide, forming a growing strand of polymerized RNA called the primary transcript. This stage of transcription is called elongation.

Recent X-ray analysis of RNA polymerase has revealed important structural details that help explain the precise mechanism of transcription. Double-helical DNA enters a long cleft in the surface of the enzyme, and is held in place by a large flexible portion of the enzyme termed the "clamp." Within the cleft, the DNA is separated and RNA is paired to it. A magnesium ion, sitting at the critical point where RNA nucleotides are added to the primary transcript, is thought to help **catalyze** this reaction. An internal barrier forces a bend in the growing DNA-RNA duplex, exposing the RNA end for addition of the incoming nucleotide. A short protein extension, termed the "rudder," helps to separate the RNA from the DNA, and the two exit the polymerase along separate paths.

The average maximum rate of elongation in bacteria is 5 to 10 nucleotides per second. However, during transcription, the polymerase enzyme may pause for seconds to minutes. These pauses are thought to be part of a regulatory mechanism. Transcription continues until RNA polymerase reaches a special DNA sequence called the termination sequence, at which point it detaches from the DNA. SEE ALSO NUCLEOTIDE; RNA; RNA PROCESSING; TRANSCRIPTION; TRANSCRIPTION FACTORS.

Richard Robinson

transcription factors proteins that increase the rate of gene transcription

catalyze aid in the reaction of

Bibliography

Alberts, Bruce, et al. *Molecular Biology of the Cell*, 3rd ed. New York: Garland Publishing, 1994.

Klug, Aaron. "A Marvellous Machine for Making Messages." *Science* 292, no. 5523 (2001): 1844–1846.

White, Robert J. *Gene Transcription: Mechanisms and Control.* Oxford: Blackwell Science, 2001.

Internet Resource

"Transcribed DNA." Euchromatin Forums. <http://www.euchromatin/org/1844-1-med.gif>.

RNA Processing

RNA serves a multitude of functions within cells. These functions are primarily involved in converting the genetic information contained in a cell's DNA into the proteins that determine the cell's structure and function. All RNAs are originally transcribed from DNA by RNA polymerases, which are specialized **enzyme** complexes, but most RNAs must be further modified or processed before they can carry out their roles. Thus, RNA processing refers to any modification made to RNA between its transcription and its final function in the cell. These processing steps include the removal of extra sections of RNA, specific modifications of RNA bases, and modifications of the ends of the RNA.

enzyme a protein that controls a reaction in a cell

Types of RNA

There are different types of RNA, each of which plays a specific role, including specifying the amino acid sequence of proteins (performed by messenger RNAs, or mRNAs), organizing and catalyzing the synthesis of proteins (ribosomal RNAs or rRNAs), translating codons in the mRNA into amino acids (transfer RNAs or tRNAs) and directing many of the RNA processing steps (performed by small RNAs in the nucleus, called snRNAs and snoRNAs).

All of these types of RNAs begin as primary transcripts copied from DNA by one of the RNA polymerases. One of the features that separates **eukaryotes** and **prokaryotes** is that eukaryotes isolate their DNA inside a nucleus while protein synthesis takes place in the cytoplasm. This separates the processes of transcription and translation in space and time. Prokaryotes, which lack a nucleus, can translate an mRNA as soon as it is transcribed by RNA polymerase. As a consequence, there is very little processing of prokaryotic mRNAs. By contrast, in eukaryotic cells many processing steps occur between mRNA transcription and translation. Unlike the case of mRNAs, both eukaryotes and prokaryotes process their rRNAs and tRNAs in broadly similar ways.

eukaryotes organisms with cells possessing a nucleus

prokaryotes single-celled organisms without a nucleus

Types of RNA Processing

There are three main types of RNA processing events: trimming one or both of the ends of the primary transcript to the mature RNA length; removing internal RNA sequences by a process called RNA splicing; and modifying RNA nucleotides either at the ends of an RNA or within the body of the RNA. We will briefly examine each of these and then discuss how they are applied to the various types of cellular RNAs.

Almost all RNAs have extra sequences at one or both ends of the primary transcripts that must be removed. The removal of individual nucleotides from the ends of the RNA strand is carried out by any of several ribonucleases (enzymes that cut RNA), called exoribonucleases. An entire section of RNA sequence can be removed by cleavage in the middle of an RNA strand. The enzymes responsible for the cleavage in this location are called endoribonucleases. Each of these ribonucleases is targeted so that it only cleaves particular RNAs at particular places.

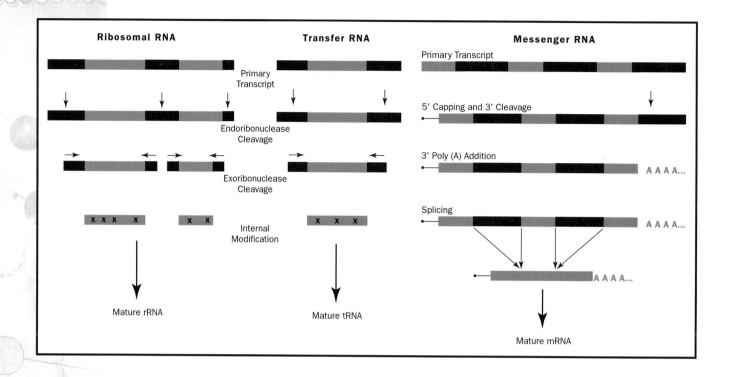

Ribosomal RNA

Primary Transcript

Endoribonuclease Cleavage

Exoribonuclease Cleavage

X X X X X   X X

Internal Modification

Mature rRNA

Transfer RNA

X X X

Mature tRNA

Messenger RNA

Primary Transcript

5' Capping and 3' Cleavage

3' Poly (A) Addition   A A A A...

Splicing   A A A A...

A A A A...

Mature mRNA

The principal steps in processing various types of RNAs in eukaryotic nuclei. RNA sequences that are in the mature RNAs are indicated in blue while RNA sequences that are removed during processing are in black. Small "x's" indicate sites of internal modifications of RNAs. The various steps do not necessarily occur in the order shown.

RNA splicing is similar to trimming in that it removes extra RNA sequences, but it is different because the sequence is removed from the middle of an RNA and the two flanking pieces are joined together again (see figure). The part of the RNA that is removed is called an intron, whereas the two pieces that are joined together, or spliced, are called exons. Just as with the cleavage enzymes, the splicing machinery recognizes particular sites within the RNA, in this case the junctions between exons and introns, and cleaves and rejoins the RNA at those positions.

Modification of RNA nucleotides can occur at the ends of an RNA molecule or at internal positions. Modification of the ends can protect the RNA from degradation by exoribonucleases and can also act as a signal to guide the transport of the molecule to a particular subcellular compartment. Some internal modifications, particularly of tRNAs and rRNAs, are necessary for these RNAs to carry out their functions in protein synthesis. Some internal modifications of mRNAs change the sequence of the message and so change the amino acid sequence of the protein coded for by the mRNA. This process is called RNA editing. As with the other types of RNA processing, the enzymes that modify RNAs are directed to specific sites on the RNA.

Processing of Various Classes of RNAs

Ribosomal RNAs are synthesized as long primary transcripts that contain several different rRNAs separated by spacer regions (see figure). The individual rRNAs are cut apart by endoribonucleases that cleave within the spacer regions. Other enzymes then trim the ends to their final length. Ribosomal RNAs are also modified at many specific sites within the RNA. Ribosomal RNA synthesis and processing occurs in a special structure within the nucleus called the **nucleolus**. The mature rRNAs bind to ribosomal proteins within the nucleolus and the assembled ribosomes are then transported to the **cytoplasm** to carry out protein synthesis.

nucleolus portion of the nucleus in which ribosomes are made

cytoplasm the material in a cell, excluding the nucleus

Transfer RNAs are transcribed individually from tRNA genes. The primary transcripts are trimmed at both the 5′ and 3′ ("five prime," or "upstream" and "three prime," or "downstream") ends, and several modifications are made to internal bases. Many eukaryotic tRNAs also contain an intron, which must be removed by RNA splicing. The finished tRNAs are then transported from the nucleus to the cytoplasm.

Messenger RNAs are transcribed individually from their genes as very long primary transcripts. This is because most eukaryotic genes are divided into many exons separated by introns. Genes may contain from zero to more than sixty introns, with a typical gene having around ten. Introns are spliced out of primary RNA transcripts by a large structure called the **spliceosome**. The spliceosome does not move along the RNA but is assembled around each intron where it cuts and joins the RNA to remove the intron and connect the exons. This must be done many times on a typical primary transcript to produce the mature mRNA.

In addition to removal of the introns, the mRNA is modified at the 5′ end by the addition of a special "cap" structure that is later recognized by the translation machinery. The mRNA is also trimmed at the 3′ end and several hundred adenosine nucleotides are added. This modification, which is called either polyadenylation or poly (A) addition, helps stabilize the 3′ end against degradation and is also recognized by the translation machinery. Finally, the processed mature mRNA is transported from the nucleus to the cytoplasm.

Some RNAs, called small nuclear RNAs (snRNAs) and small nucleolar RNAs (snoRNAs), are processed in the nucleus and are themselves part of the RNA processing systems in the nucleus. Most snRNAs are involved in mRNA splicing, while most snoRNAs are involved in rRNA cleavage and modification.

spliceosome RNA-protein complex that removes introns from RNA transcripts

RNA Processing and the Human Genome

The fact that most human genes are composed of many exons has some important consequences for the expression of genetic information. First, we now know that many genes are spliced in more than one way, a phenomenon known as alternative splicing. For example, some types of cells might leave out an exon from the final mRNA that is left in by other types of cells, giving it a slightly different function. This means that a single gene can code for more than one protein. Some complicated genes appear to be spliced to give hundreds of alternative forms. Alternative splicing, therefore, can increase the coding capacity of the genome without increasing the number of genes.

A second consequence of the exon/intron gene structure is that many human gene mutations affect the splicing pattern of that gene. For example, a mutation in the sequence at an intron/exon junction that is recognized by the spliceosome can cause the junction to be ignored. This causes splicing to occur to the next exon in line, leaving out the exon next to the mutation. This is called exon skipping and it usually results in an mRNA that codes for a nonfunctional protein. Exon skipping and other errors in splicing are seen in many human genetic diseases. SEE ALSO ALTERNATIVE SPLICING; NUCLEOTIDE; NUCLEUS; RIBOSOME; RNA; RNA POLYMERASES.

Richard A. Padgett

Bibliography

Lewin, Benjamin. *Genes VII.* Oxford, U.K.: Oxford University Press, 2000.

Lodish, Harvey, et al. *Molecular Cell Biology*, 4th ed. New York: W. H. Freeman, 2000.

Rodent Models

Rodents play an important role in biology and medicine. Since the mouse and rat share many biological characteristics with humans, they are commonly used as model organisms for understanding disease processes and testing treatments. Moreover, it is relatively easy to experimentally manipulate the genetic composition of mice and rats and, thereby, to model human genetic disorders in these animals.

Large numbers of mice can be raised quickly and relatively cost-effectively. Mice have short life spans and relatively short generation times. This makes the mouse an extremely useful experimental system for studying the genetics and biology of human disease. Rats are slightly more similar to humans than are mice, but rats are larger, have a longer generation time, are less well-characterized genetically, and are not as easily genetically manipulated. So while rats are useful in the study of some aspects of biology, mice are now much more widely used in molecular genetic research.

Many features of human biology at the cellular and molecular levels are shared with a wide variety of organisms. As mammals, the mouse and rat are highly related to humans, with similar genes, biochemical pathways, organs, and physiology.

genomes total genetic material in a cell or organism

The **genomes** of both human and mouse have now been sequenced, and they show striking similarity. Estimates of the total number of genes in both genomes are very comparable, ranging from 30,000 to 40,000. It is thought that only about 1 percent of human genes are unique and do not have a mouse counterpart. In addition, mouse genes are on average approximately 85 percent similar in sequence to their human counterparts. Thus, mice and humans are very highly related at the genetic level.

mutations changes in DNA sequences

Before the advent of genetic modification techniques in the twentieth century, collecting mouse strains with unusual characteristics was a pastime of amateur enthusiasts. These variant strains, often showing physical characteristics such as unusual fur or tails, arose as a result of spontaneous **mutations**. The discovery that radiation and chemicals could increase the rate of mutation led to the development of laboratory mouse stocks, which are a useful source of genetic variation. Over 1,000 spontaneous or radiation-induced mutations have been documented, and many have been the starting point for the study of relevant human diseases and biology. For example, the cloning of the obese and diabetic mouse mutants identified the **hormone** leptin and its receptor, respectively, and opened up an entirely new area of research into the control of various physiological processes, including the control of body weight.

hormones molecules released by one cell to influence another

The importance of mice in genetic studies was first recognized in the related areas of immunology and cancer research, for which a mammalian model of human diseases was essential. Mice have been used in cancer research for almost a century, starting with the breeding of mouse strains

Humans share many biological characters with rats and mice, which makes the rodents valuable models in the laboratory for researchers who are looking to find cures for genetic diseases.

susceptible to particular tumor types and, subsequently, in the induction of tumors by chemical compounds. These experiments were critical for demonstrating that the development of cancer is a multistep process, consisting of a series of genetic changes.

Although it has long been obvious that many other aspects of human biology and development can be studied using mouse models, until recently the methods for doing so did not exist. Mouse models moved to the forefront of modern biomedical research with the emergence of **recombinant DNA** technology, thirty years ago, and the pace has been accelerating ever since. The advent of this technology eventually culminated in the publication of the complete human genome sequence in 2001, and by 2002 the sequence and structure of almost every mouse gene will also be known. In combination with advances in DNA cloning, methods were developed for adding genes to mice (called transgenes) and removing genes (called gene targeting). These introduced genetic changes allow researchers to model human diseases by altering specific genes that are altered in particular human disease states.

recombinant DNA DNA formed by combining segments of DNA, usually from different types of organisms

61

For example, the mouse counterpart of *atm*, the gene that causes the human disease ataxia telangiectasia, was identified. In humans the symptoms of this disease include movement problems, abnormal blood vessels, immune defects, and a predisposition to cancer. The mouse *atm* gene was specifically disrupted by gene targeting, and mice lacking this gene exhibit a range of problems similar to those exhibited by the human patients. This mouse model is useful not only for studying and understanding the function of the gene in humans and its role in the disease process, but also for testing potential therapies.

Now that the sequencing of the human and mouse genes is completed, there is a large gap between the number of genes that have been identified and our understanding of how those genes function. The next challenge is to start to determine how those genes function in normal development and how these genes are disrupted in various disease conditions. Arguably, mouse models will become even more important during this phase of the analysis of the genome, as mice are amenable to a systematic genome-wide mutation of genes, which will allow the function of large numbers of genes to be determined. SEE ALSO GENE TARGETING; MODEL ORGANISMS; RECOMBINANT DNA; TRANSGENIC ANIMALS.

Seth G. N. Grant and Douglas J. C. Strathdee

Bibliography

Jackson, Ian J., and Catherine M. Abbot, eds. *Mouse Genetics and Transgenics: A Practical Approach.* New York: Oxford University Press, 2000.

Lyon, Mary F., Sohaila Rastan, and S. D. M. Brown. *Genetic Variants and Strains of the Laboratory Mouse,* 3rd ed. New York: Oxford University Press, 1996.

Rossant, Janet, and Patrick P. L. Tam, eds. *Mouse Development: Patterning, Morphogenesis, and Organogenesis.* San Diego: Academic Press, 2001.

Roundworm: *Caenorhabditis elegans*

nematode worm of the Nematoda phylum, many of which are parasitic

Caenorhabditis elegans is a nonparasitic **nematode** that normally lives in the soil. Although studied since the 1800s, the modern use of *C. elegans* as a model system dates to the mid-1960s. By the start of that decade, scientists thought that the "classical" problems in molecular biology were about to be solved. They began to search for a model system that would support the new challenges presented by research on multicellular organisms, particularly in developmental biology and neurology. In 1965 Sydney Brenner proposed what he thought was the ideal model organism, one that provided the best compromise between biological complexity and ease of manipulation: *C. elegans*.

Useful Characteristics

morphologically related to shape and form

mutations changes in DNA sequences

C. elegans is hermaphroditic, meaning that almost all of the 300 progeny produced in a single clutch of eggs are females, capable of self-fertilization. In essence, *C. elegans* clones itself. One or two progeny are **morphologically** distinct males that can be collected and used for standard genetic studies. This permits rapid chromosomal mapping of the many **mutations** that occur in any one of the organism's six chromosomes. The *C. elegans* genome was sequenced in 1998, predicting 19,099 genes. Currently, 3,000 of these are considered

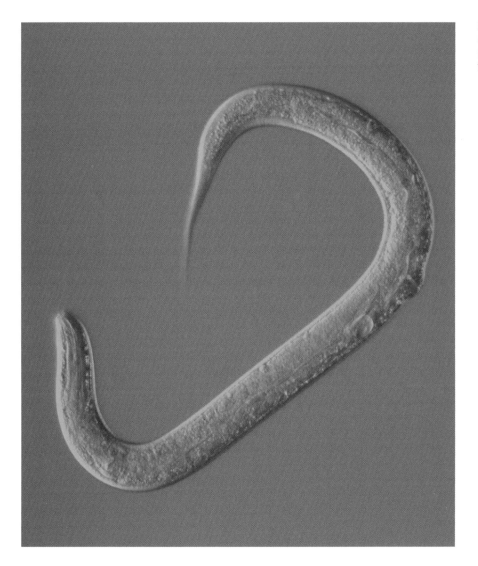

Micrograph of a
Caenorhabditis elegans
nematode, magnified fifty
times.

"essential" genes, 900 of which had already been identified through chromosomal mapping.

In the laboratory, *C. elegans* is easily cultivated on simple growth **media**
on petri dishes. Only 1 millimeter long, its transparent nature permits an inves-
tigator to trace the origin and determine the exact anatomical position of any
one of its 959 cells at any stage in development from fertilized egg to death.
This has become known as "cell lineage." Cell abnormalities, as well as abnormal migration of normal cells, are readily discovered, which has led to the identification and cataloging of a large number of mutants. The ability to directly
see the individual sarcomeres (protein subunits) within a muscle cell quickly
established *C. elegans* as a model system for the study of muscle structure and
function. The gene for the muscle protein myosin, which could not be identified or cloned in other organisms, was readily cloned from *C. elegans*. This
accomplishment, along with the study of the genes that give rise to "paralyzed"
mutants of *C. elegans*, has had a major impact on the study of muscle development and diseases in humans. The transparency of *C. elegans* continues to
expand the universality of this model system through the use of techniques that
were not available in 1965, such as microinjection or laser microsurgery.

media nutrient source

somatic nonreproductive; not an egg or sperm

apoptosis programmed cell death

Insights from Laboratory Observation

Although anatomically and genetically simple, *C. elegans* mimics the life cycle of humans. Starting from a fertilized egg, it undergoes a complex development that gives rise to excretory, reproductive, digestive, and neuromuscular organ systems. The cell lineages of each of the 959 adult **somatic** cells have been directly observed, as has the fact that an additional unique set of 131 cells die during development. These deaths would have gone unnoticed, except for their demise was observable under the microscope. These observations led to the concept that **apoptosis** is a vital feature of development in multicellular organisms, and that it enables the shaping and carving of organs and tissues.

Today *C. elegans* remains one of the foremost model systems used in genetic research. Advanced descriptions have been made of its anatomy, cell lineages, developmental genetics, neural development, and reproductive cycle. The recent discovery of a mutation that doubles its life expectancy suggests that this "worm" will continue to expand its usefulness as a model system for the study of aging. Its growing significance as a member of the large, evolutionarily successful Nematoda phylum highlights the importance of *C. elegans* to the study of the genes involved in adaptive evolution. SEE ALSO APOPTOSIS; CLONING GENES; GENOME; MAPPING; MODEL ORGANISMS.

Diane C. Rein

Bibliography

Alberts, B., et al. *Molecular Biology of the Cell*, 3rd ed. New York: Garland Publishing, 1994.

Riddle, D. L., et al., eds. *C. elegans II*. Plainfield, NY: Cold Spring Harbor Laboratory Press, 1997.

Roberts, Leslie. "The Worm Project." *Science* 248, no. 4962 (1990): 1310–1313.

Sanger, Fred

Molecular Biologist
1918–

Born August 13, 1918, in Rendcombe, Gloucestershire, United Kingdom, Fred Sanger has been breaking new ground in chemistry for decades. In fact, he is the only person to have won a Nobel Prize in chemistry twice, and is only one of four people ever to have won a Nobel Prize more than once.

While at Cambridge University in England he developed a new method for sequencing amino acids in proteins, which he used to identify the complete sequence of insulin. For this he was awarded his first Nobel Prize in chemistry in 1958. After this success, in 1961 Sanger moved to the MRC Laboratory of Molecular Biology, where he became head of the division of protein chemistry. His colleagues' interest in nucleic acids inspired him to turn his interest in sequencing to the research of nucleic acids.

In 1977 Sanger developed a sequencing method, called the "dideoxy" method, with which he determined the entire sequence of a bacterial virus called phi-X174. This was the first time a complete sequence of a DNA molecule had been established. For this achievement he was awarded the

1980 Nobel Prize in chemistry, shared jointly with Walter Gilbert of Harvard University who had concurrently developed an alternative DNA sequencing method. Sanger's original sequence contained only 5,375 nucleotides, but his technology is now being used to determine sequences that are millions of nucleotides longer, including, importantly, the human genome. In his honor one of the major DNA sequencing centers in the world is named the Wellcome Trust Sanger Institute, at the Wellcome Trust Genome Campus, Cambridge, England. SEE ALSO SEQUENCING DNA.

Jeffery M. Vance

Bibliography

Internet Resource

"About Frederick Sanger." Wellcome Trust Sanger Institute. <http://www.sanger.ac.uk/Info/Intro/sanger.shtml>.

Fred Sanger shortly after receiving his second Nobel Prize for chemistry.

Science Writer

All good journalists are storytellers, and science writers are simply journalists who like to tell stories about science. Still, it has only been in the last few decades that science writing has become a profession in its own right, with journalists specifically trained to cover research ranging from genetics to particle physics.

The Growth of Specialization

Previously, most publications put so little emphasis on science that they did not need a specialized reporter. More and more, though, news organizations regard scientific results as necessary information, part of the everyday reporting of news. After all, science and technology have radically altered the way we live. For example, consider the development of antibiotics and vaccines, nuclear weapons, computer technologies, lasers, and fiber optics. Advances in science and technology will undoubtedly continue to occur in ways that we can not fully predict. The Human Genome Project, with all its promise and ethical unknowns, illustrates this perfectly.

The fast pace of scientific and technological advance has led to an increasing demand for science writers who understand science, and who can make others understand it as well. The membership of the National Association of Science Writers is now nearly 2,500. Science writing programs have sprung up at Boston University, Northwestern University, the University of California in Santa Cruz, the University of Maryland, and many other institutions. Most of these programs are aimed at journalists who wish to learn how to write about science, to more deftly translate jargon, explain complex experiments, and illuminate the people and the politics behind the science. A few, such as the program at Santa Cruz, are geared for science majors who wish to learn about journalism.

Breaking into the Field

While lack of science training is not a bar to working as a science journalist, a growing number of working science writers have undergraduate science degrees, or a combined degree in science and journalism. The bigger

publications and broadcast organizations place a premium on good writers with scientific training. Beyond that, being a trained researcher enables a journalist to ask questions that a colleague less knowledgeable might miss. Writers who do more specialized types of writing, such as medical writing for a physician audience, especially benefit from background in their field. Very few working science writers have advanced degrees, but there are advantages, both for the skills and knowledge gained and the credentials, which may lead to opportunities not otherwise available.

While valuing a scientific background, many media organizations put an even higher premium on good writing and the ability to make a difficult subject accessible to nonspecialist readers. A talented and hardworking journalist who is willing to do the necessary research and homework is thus often a valuable commodity even without specific science training.

Opportunities in Science Journalism

Opportunities for science writers continue to expand beyond such traditional media as newspapers, magazines, television, and radio. For example, there are on-line publications, some of which specialize only in research topics. Science writers are also hired as public information officers, to explain research at universities and government agencies. There are comparable jobs in private industry, working for specific companies or trade publications. Many of the big science associations publish their own journals or magazines and hire science-trained journalists for their news and comment sections.

The latter jobs mean writing for a science-literate audience. For the most part, however, science journalists explain a technical world to a nontechnical audience. Their goal is not to teach people how to do the science, just how to appreciate it, evaluate it, and even enjoy it. Continually learning new science, and explaining it to those who are interested, is one of the great benefits of being a science writer.

Compensation

Beginning science writers may earn a starting salary of around $20,000 to $25,000, depending on their training, the type of job, the resources of the employer, and the region of the country. More experienced writers may earn between $35,000 and $60,000, and some writers earn even more. The highest salaries are paid to top-tier writers employed by major publications, or experienced medical writers working for pharmaceutical companies. SEE ALSO TECHNICAL WRITER.

Deborah Blum

Bibliography

Blum, Deborah, and Mary Knudson, eds. *A Field Guide for Science Writers.* New York: Oxford University Press, 1998.

Friedman, Sharon M., Sharon Dunwoody, and Carol L. Rogers, eds. *Communicating Uncertainty: Media Coverage of New and Controversial Science.* Mahwah, NJ: Lawrence Erlbaum Associates, 1999.

Gastel, Barbara. *The Health Writers Handbook.* Ames: Iowa State University Press, 1998.

Internet Resources

American Association for the Advancement of Science: Media Relations. <http://www.eurekalert.org>.

National Association of Science Writers. <http://www.nasw.org>.

Society of Environmental Journalists. <http://www.sej.org>.

Selection

Selection is a process in which members of a population reproduce at different rates, due to either natural or human-influenced factors. The result of selection is that some characteristic is found in increasing numbers of organisms within the population as time goes on.

Types of Selection

Artificial selection, which is even older than agriculture, refers to a conscious effort to use for future breeding those varieties of a plant or animal that are most useful, attractive, or interesting to the breeder. Artificial selection is responsible for creating the enormous number of breeds of domestic dogs, for instance, as well as high-yielding varieties of corn and other agricultural crops.

Selection also occurs in nature, but it is not conscious. Charles Darwin called this natural selection. Darwin saw that organisms constantly vary in a population from generation to generation. He proposed that some variations allow an organism to be better adapted to a given environment than others in the population, allowing them to live and reproduce while others are forced out of reproduction by death, sterility, or isolation. These genetic variations gradually replace the ones that fail to survive or to reproduce. This gradual adjustment of the **genotype** to the environment is called adaptation. Natural selection was not only Darwin's key mechanism of evolution for the origin of species, it is also the key mechanism today for understanding the evolutionary biology of organisms from viruses to humans. Natural selection leads to evolution, which is the change in gene frequencies in a population over time.

genotype set of genes present

The concept of selection plays an increasingly important role in biological theory. New fields such as evolutionary psychology rely heavily on natural selection to explain the evolution of human behavioral traits, such as mate choice, aggression, and other types of social behavior. A great difficulty in such a theoretically based science is the paucity of experimental or direct evidence for presumed past environments and presumed behavioral responses that were genetically adaptive.

Variation

The variation that selection requires arises from two distinct sources. The ultimate sources of variation are gene **mutation**, gene duplication and disruption, and chromosome rearrangements. Gene mutations are randomly occurring events that at a molecular level consist mostly of substitutions or small losses or gains of nucleotides within genes. Gene duplication makes new copies of existing genes, while gene disruptions destroy functional copies of genes, often through insertion of a mobile genetic element. Chromosome

mutation change in DNA sequence

rearrangements are much larger changes in chromosome structure, in which large pieces of chromosomes break off, join up, or invert. Individually, such mutations are rare. Most small mutations are either harmful or have no effect, and they may persist in a population for dozens or hundreds of generations before their advantages or disadvantages are evident.

The second source of variation arises from the shuffling processes undergone by genes and chromosomes during reproduction. During **meiosis**, maternally and paternally derived chromosome pairs are separated randomly, so that each sperm or egg contains a randomly chosen member of each of the twenty-three pairs. The number of possible combinations is over eight billion. Even more variation arises when pair members exchange segments before separating, in the process known as crossing over. The extraordinary variety in form exhibited even by two siblings is due primarily to the shuffling of existing genes, rather than to new mutations.

meiosis cell division that forms eggs or sperm

The Importance of the Environment

A disadvantageous trait in one environment may be advantageous in a very different environment. A classic example of this is sickle cell disease in regions where malaria is common. Individuals who inherit a copy of the sickle cell gene from both of their parents (**homozygotes**) die early from the disease, whereas heterozygotes (individuals who inherit only one copy of the gene) are favored in malarial areas (including equatorial Africa) over those without any copies, because they contract milder cases of malaria and thus are more likely to survive it.

homozygotes individuals whose genetic information contains two identical copies of a particular gene

Even though homozygotes rarely pass on their genes, because of their low likelihood of surviving to reproduce, the advantage of having one copy is high enough that natural selection continues to favor presence of the gene in these populations. Thus a malarial environment can keep the gene frequency high. However, in temperate regions where malaria is absent (such as North America), there is no heterozygote advantage to the sickle cell gene. Because heterozygotes still suffer from the disease, they are less likely to survive and reproduce. Thus, selection is gradually depleting the gene from the African American population that harbors it.

Artificial Selection

One of the first uses of genetic knowledge to improve yields and the quality of plant products was applied to **hybrid** seed production at the start of the twentieth century by George Shull. Artificial selection today is still done by hobbyists who garden or raise domestic animals. It is done on a more professional level in agriculture and animal breeding. The benefits are enormous. Virtually all commercial animal and plant breeding uses selection to isolate new combinations of traits to meet consumer needs. In these organisms, most of the variation is preexisting in the population or in related populations in the wild. The breeder's task is to combine (hybridize) the right organisms and select offspring with the desired traits.

hybrid combination of two different types

In the antibiotic industry selection is used to identify new antibiotics. Usually, microorganisms are intentionally mutated to produce variation. Mutations can be induced with a variety of physical and chemical agents called mutagens, which randomly alter genes. Some early strains of penicillin-producing molds were x-rayed and their mutations selected for higher yields.

Biologists also make use of selection in the process called molecular cloning. Here, a new gene is inserted into a host along with a marker gene. The marker is typically a gene for antibiotic resistance. To determine if the host has taken up the new genes, it is exposed to antibiotics. The ones who survive are those that took up the resistance gene, and so also have the gene of interest. This selection process allows the researcher to quickly isolate only those organisms with the new gene.

Selection in Humans

Both natural and artificial selection occur in human beings. If a trait is lethal and kills before reproductive maturity, then that gene mutation is gradually depleted from the population. Mutations with milder effects persist longer and are more common than very severe mutations, and **recessive** mutations persist for much longer than **dominant** ones. With a recessive trait, such as albinism, the parents are usually both carriers of a single copy of the gene and may not know that they carry it. If a child receives a copy of this gene from both of the carrier parents, the albino child may die young, may find it difficult to find a partner, or may end up marrying much later in life. This is usually considered a form of natural selection.

recessive requiring the presence of two alleles to control the phenotype

dominant controlling the phenotype when one allele is present

Considerable abuse of genetic knowledge in the first half of the twentieth century led to the **eugenics** movement. Advocates of eugenics claimed some people were more fit and others less fit (or unfit), and argued that the least fit should be persuaded or forced not to reproduce. Eugenicists typically defined as unfit those who were "feeble-minded, criminal, socially deviant, or otherwise undesirable." Coerced sterilization, a form of artificial selection, was practiced on some of these individuals. SEE ALSO CLONING GENES; EUGENICS; HARDY-WEINBURG EQUILIBRIUM; MULLER, HERMANN; MUTAGENESIS; MUTATION.

eugenics movement to "improve" the gene pool by selective breeding

Elof Carlson

Bibliography

Huxley, Julian. "Adaptation and Selection." In *Evolution: The Modern Synthesis*. New York: Harper and Brothers, 1942.

Pianka, Eric. *Evolutionary Biology*, 6th ed. San Francisco: Addison-Wesley-Longman, 2000.

Sequencing DNA

A gene is a segment of DNA that carries the information needed by the cell to construct a protein. Which protein that is, when it is made, and how damage to it can give rise to genetic disease all depend on the gene's sequence. In other words, they depend on how the building blocks of DNA, the **nucleotides** A, C, G, and T (adenine, cytosine, guanine, and thymine) are ordered along the DNA strand. For example, part of a gene may contain the base sequence TGGCAC, while part of another gene may contain the base sequence TCACGG. Knowing a gene's base sequence can lead to

nucleotides the building blocks of RNA or DNA

isolation of its protein product, show how individuals are related, or point the way to a cure for those people carrying it in its damaged form.

Overview

In 1977 two methods for sequencing DNA were introduced. One method, referred to as Maxam-Gilbert sequencing, after the two scientists at Harvard University who developed the technique, uses different chemicals to break radioactively labeled DNA at specific base positions. The other approach, developed by Frederick Sanger in England and called the chain termination method (also called the Sanger method), uses a DNA synthesis reaction with special forms of the four nucleotides that, when added to a DNA chain, stop (terminate) further chain growth.

By either method, a collection of single-stranded DNA fragments is produced, each fragment one base longer than the next. The length of a fragment depends on where a chemical cleaved the strand (in Maxam-Gilbert sequencing) or where a special terminator base was added (in the chain termination method). The fragments are then separated according to their size by a process called gel electrophoresis, in which the fragments are drawn through a gel material by electric current, with shorter fragments migrating through the gel faster than longer fragments. The DNA sequence is then "read" by noting which reaction produced which fragment.

Maxam-Gilbert Sequencing

All DNA sequencing protocols must incorporate a method for making the DNA fragments generated in the reaction "visible"; they must be capable of being detected. For Maxam-Gilbert sequencing a technique called end labeling is used, in which a radioactive atom is added to the ends of the DNA fragment being sequenced. The first step in this process is to use an **enzyme**, called a restriction endonuclease, to cut the DNA at a specific sequence. If the restriction endonuclease is Hind III, for example, the sequence AAGCTT will be cut.

The ends of the DNA segment made by the restriction endonuclease will have **phosphate groups** ($-PO_3^{2-}$) at its ends. An enzyme called a phosphatase is used to remove the phosphate group. Another enzyme, called a kinase, is then used to add a radioactive phosphate in its place. This reaction will add radioactive atoms onto both ends of the DNA restriction fragment. A second restriction endonuclease is used to make a cut within the end-labeled fragment and **gel electrophoresis** is then employed to separate the two resulting subfragments from each other. Each subfragment now has one labeled and one unlabeled end. The subfragment whose sequence is to be determined is cut out of the gel to purify it away from the other end-labeled subfragment.

The end-labeled piece of DNA is then divided and the fragments are placed in four separate tubes. They are then treated with different chemicals that weaken and break the bond holding the base to the backbone of the DNA molecule. These chemicals are base specific. In other words, one chemical causes the "C" reaction, in which the bond holding the C base in position is broken. Another chemical breaks the bond holding the G in place (the "G" reaction). Another breaks both G and A bases from the DNA back-

enzyme a protein that controls a reaction in a cell

phosphate groups PO_4 groups, whose presence or absence often regulates protein action

gel electrophoresis technique for separation of molecules based on size and charge

bone (the "G+A" reaction), and a fourth breaks the bonds holding the C and T bases in place (the "C+T" reaction).

Each reaction is limited so that each DNA molecule will have only one of its base positions altered. Within the "C" reaction, for example, each DNA molecule will have only one of its C bases weakened and removed. One of the DNA molecules may have the C base closest to the end dislodged. On another DNA molecule, a C that appears 500 bases away from the end may be removed. Every C position in the population of molecules, however, is subject to treatment.

When the first step in the reaction is concluded, another reagent is then used to completely break the DNA strands at the points where the bases have been removed. In this way, a collection of DNA fragments is generated that differ in size according to the position along the strand where the break occurred.

Each reaction is electrophoresed in its own lane on a gel that separates the DNA fragments by their length. A process called autoradiography is then used to detect the separated fragments. In this technique, X-ray sensitive film is placed flat against the gel under conditions of complete darkness. Since the fragments made in the sequencing reactions are end-labeled with radioactive atoms, their emissions will expose the X-ray film at the positions where they are found on the gel. When the X-ray film is developed, bands, like a bar code, reveal the sizes of the fragments generated in each separate reaction.

Each band that has been rendered visible on the X-ray film differs from the one above or the one below it by a single base. The DNA sequence is then read from the bottom of the gel upward. A band found in the "G" reaction lane is read as a G, a band found in the "C" lane is read as a C, a band found in the lane from the "G+A" reaction but with no corresponding band of the same length in the "G" lane is read as an A, and a band in the "C+T" reaction without a corresponding base in the "C" lane is read as a T (see Figure 1).

Chain Termination Method

Because it employs fewer steps and does not require the use of **restriction enzymes**, the chain termination method of DNA sequencing is used in more laboratories than is the Maxam-Gilbert approach. Chain termination sequencing is a clever variation of the reaction used to replicate DNA, and requires only a handful of components. Four reaction tubes, designated "A," "C," "G," and "T," are prepared and the DNA strand whose sequence is to be determined is added into each tube. This DNA strand is called the template.

Along with the template, a short, single-stranded piece of DNA (called a primer) is added; it attaches specifically to one section of the template and serves as a starting point for synthesis of a new DNA strand. Also added are the four "building block" nucleotides, a **buffer** (to maintain the proper pH level), the enzyme DNA polymerase, and its cofactor magnesium, all of which are needed to extend the primer into a full-length DNA chain.

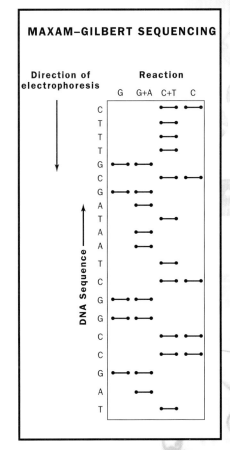

Figure 1. Schematic diagram of autoradiogram showing bands generated by the Maxam-Gilbert method of sequencing DNA. The DNA sequence is read from the bottom of the gel by noting which gel lane contains a band.

restriction enzymes enzymes that cut DNA at a particular sequence

buffers substances that counteract rapid or wide pH changes in a solution

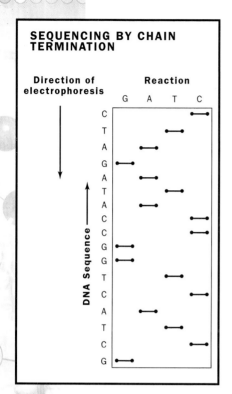

SEQUENCING BY CHAIN TERMINATION

Figure 2. Schematic diagram of autoradiogram showing bands generated by the chain termination method of DNA sequencing. A band in any particular gel lane results from the incorporation of a dideoxynucleotide. DNA sequence is read from the bottom of the gel upward.

To make the DNA strands visible, a nucleotide carrying a radioactive phosphorous or sulfur atom is included in the reaction. Also included in each reaction tube is a quantity of a unique dideoxynucleotide, a modified form of a nucleotide that lacks the site at which other bases can attach during chain growth. Thus, when this nucleotide is added to the DNA chain, all further chain growth is terminated. In the "A" tube, a dideoxynucleotide form of the A base, dideoxyadenosine triphosphate, is added. In the "C" tube, dideoxycytidine triphosphate is added. Dideoxyguanosine triphosphate is added to the "G" tube, and dideoxythymidine triphosphate to the "T" tube.

When the reactions are incubated at a temperature suitable for the DNA polymerase, nucleotides that are complementary to the bases on the template are added onto the end of the attached primer. The bases A and T are complementary, as are G and C. Thus, if a T base is on the template strand, DNA polymerase will add the complementary base, A, at the corresponding position in the new, extending strand. If, by chance, the DNA polymerase adds a dideoxynucleotide, chain growth is terminated at that point.

Since dideoxynucleotides are randomly incorporated, each reaction generates a mixture of DNA molecules of variable length, but all are terminated by a dideoxynucleotide. Careful adjustment of reactant concentration will give a set of DNA molecules that terminate at each of the possible positions. The "A" tube, for instance, will contain a mixture of DNA chains, each of which ends in a different "A."

As in Maxam-Gilbert sequencing, each reaction is loaded in its own gel lane and electrophoresal autoradiography is then used to detect the fragments. The base sequence is read directly off the X-ray film from the bottom of the gel upward, noting the lane in which each band appears (see Figure 2). The bands at the bottom of the gel represent the shortest fragments and resulted from termination events closest to the primer. Bands toward the top of the gel represent longer fragments made by termination events farthest from the primer.

Automated Sequencing with Fluorescent Dyes

As originally developed, both the Maxam-Gilbert and the chain termination methods of DNA sequencing require the use of radioisotopes to visualize the fragments generated in the reactions. However, in addition to being a health risk and presenting a disposal problem, the use of radioactivity makes automation of the DNA sequencing process difficult. Machines that could automatically read DNA sequences did not become practical until fluorescent dyes were introduced as a way to label sequencing fragments.

Two approaches are used to fluorescently label the products of a DNA sequencing reaction: the dye primer method and the dye terminator method. Both are applied only to the chain termination method. The sequencing products made by these methods are electrophoresed on an instrument that uses a laser to detect the different fragments.

In the dye primer method, fluorescent dyes are attached to the 5′ ("five prime") end of the primer, which is the end opposite to that where nucleotides are added during chain growth. Four reactions are prepared con-

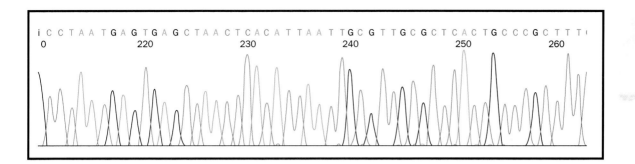

i C C T A A T G A G T G A G C T A A C T C A C A T T A A T T G C G T T G C G C T C A C T G C C C G C T T T

0   220   230   240   250   260

Figure 3. Chromatogram produced by an Applied Biosystems automated DNA sequencer. DNA sequence is read by a computer based on the fluorescent dye color associated with each band (represented as a peak) on the gel.

taining all the components described above, but with no radioisotopes. In the "A" reaction, the primer carries a dye at one end that fluoresces green when struck with a laser. In the "C" reaction, the primer carries a dye that fluoresces blue. The "G" reaction's primer has a yellow dye on its end and the "T" reaction's primer carries a red dye.

Termination events in the "A" reaction tube result in fragments with a green dye at one end and a dideoxyadenosine at the other. Termination events in the "C" reaction result in the formation of fragments having a blue dye at one end and a dideoxycytidine at the other. Similar relationships hold in the "G" and "T" reactions. When incubation is complete, the four reactions are combined and electrophoresed in a single gel lane of an automated sequencer. A laser, shining at the bottom of the gel, excites the dyes on the DNA fragments, causing them to fluoresce as they pass. The instrument's optics system detects the fluorescent colors during electrophoresis and a computer then translates the order of colors into a base sequence (see Figure 3).

In the second approach to fluorescent DNA sequencing, the dye terminator method, dyes are attached to the dideoxynucleotides instead of to the primers. The DNA fragment in the sequencing reaction becomes dye-labeled when a dideoxynucleotide is incorporated. The dye terminator method uses a single reaction tube (rather than four) because each dideoxynucleotide is associated with a different dye. The cost and time for sequencing are therefore reduced, making this approach the preferred method used by most laboratories.

Manufacturers of DNA sequencing reagents now provide kits that contain all the components necessary for a sequencing reaction in a "master mix" format. For the dye terminator approach, a master mix can be purchased that contains DNA polymerase, the four nucleotides, buffer, magnesium, and the four dye-labeled dideoxynucleotides. The addition of **template** and primer to the master mix completes the reaction.

template a master copy

In addition to the widespread use of sequencing kits, other improvements have been made. Enhanced signal strength and improved sensitivity have been achieved through the development of stronger fluorescent dyes and the exploitation of heat-stable DNA polymerases that allow for repeated cycling of the sequencing reaction. SEE ALSO AUTOMATED SEQUENCER; CYCLE SEQUENCING; DNA; DNA POLYMERASES; GEL ELECTROPHORESIS; HUMAN GENOME PROJECT; NUCLEOTIDE; RESTRICTION ENZYMES.

Frank H. Stephenson and Maria Cristina Abilock

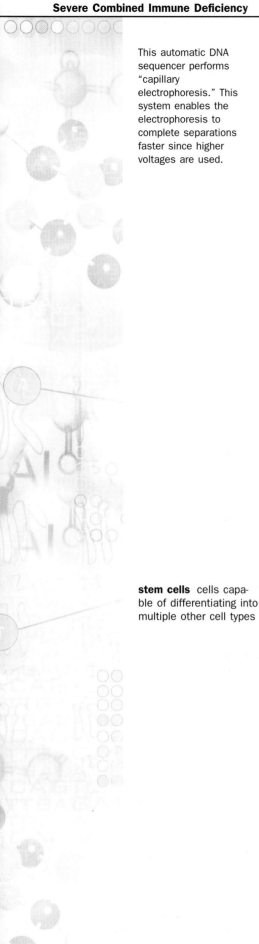

This automatic DNA sequencer performs "capillary electrophoresis." This system enables the electrophoresis to complete separations faster since higher voltages are used.

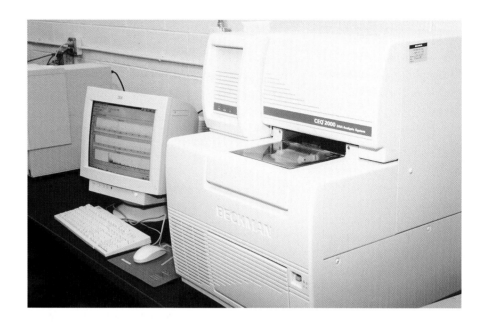

Bibliography

Bloom, Mark V., et al. *Laboratory DNA Science: An Introduction to Recombinant DNA Techniques and Methods of Genome Analysis.* San Francisco: Benjamin Cummings Publishing, 1995.

Maxam, Allan M., and Walter Gilbert. "A New Method for Sequencing DNA." *Proceedings of the National Academy of Sciences* 74 (1977): 560–564.

Sanger, Frederick, S. Nicklen, and Alan R. Coulson. "DNA Sequencing with Chain-terminating Inhibitors." *Proceedings of the National Academy of Sciences* 74 (1977): 5463–5467.

Severe Combined Immune Deficiency

stem cells cells capable of differentiating into multiple other cell types

The development of the immune system is a very complicated process. **Stem cells** in the bone marrow continually give rise to the white blood cells responsible for producing antibodies (B-lymphocytes), recognizing and destroying foreign cells (T-lymphocytes) and performing immune surveillance for cancer and foreign cells. Immunodeficiency results when the normal complex interactions of the immune system required to do this fail, resulting in susceptibility to infections or cancer.

Types and Severity of Immunodeficiency Diseases

Disease severity can range from mild to fatal, depending upon what part of the immune system is affected. Immunodeficiency can originate in normal individuals as a consequence of chemotherapy, viral infections (such as AIDS, which is caused by the HIV virus), or as the result of other processes that prevent immune system function. When immunodeficiency occurs in this manner, it is called acquired.

In contrast, immunodeficiency can also be inherited as a genetic mutation that prevents the normal development and function of the immune system. This is called primary immunodeficiency, of which there are three subtypes: mutations that prevent the function of B-lymphocytes (antibody production), those that prevent the function of T-lymphocytes ("invader" recognition), and those that affect both B- and T-lymphocyte production. The last group is called severe combined immunodeficiency disease (SCID). These patients make none or very few T-lymphocytes, have nonfunctional B-lymphocytes, and may or may not have a type of immune cell called natural killer cells. This combination results in the absence of a functioning immune system from the moment of birth.

SCID

SCID is a collection of rare diseases, estimated to occur once in every 80,000 live births. If left untreated it always results in fatal infections within the first two years of life. At first, to prolong life, patients were placed in sterile isolators, away from direct human contact. This gave rise to the name "bubble babies." Currently bone marrow transplantation is available with varying degrees of success. Gene therapy to treat SCID is under development.

Although the first known description of SCID was in 1950, very little progress in understanding the genetic basis for the disease was made until the mid-1990s. SCID patients have a wide range of symptoms that make it difficult to define the number of genes involved. The development of strains of mice that exhibited SCID, either naturally or created by a laboratory technique called gene knockout, greatly enhanced the ability to study human SCID genes and permitted the development of strategies for treatment, such as bone marrow transplantation and gene therapy. Many SCID patients were able to live longer through bone marrow transplants. The study of these patients, combined with the advances from the Human Genome Project, has led to a rapid increase in our knowledge about the genetic cause of SCID. The genes involved in five different SCID diseases were confirmed in late 2001 and are described below.

SCID-X1 (XSCID, XL-SCID). This gene is currently the only known X-linked version of SCID (all other SCID forms identified are **autosomal** and recessive). SCID-X1 accounts for 46 percent of all SCID cases and exhibits a high spontaneous mutation rate. It is caused by mutations in the gene for the γ subunit of the interleukin 2 (IL-2) **cytokine** receptor. This receptor is part of a critical cytokine signal pathway required early in immune system growth and differentiation. The most famous SCID-X1 patient was David Vetter. Known as the "bubble boy," he lived for twelve years in an isolated environment before dying from an Epstein-Barr virus infection.

ADA-SCID. This occurs in 15 percent of SCID patients. It is due to mutation in the *ADA* gene on chromosome 20. In the absence of ADA enzyme, accumulation of deoxyATP (a DNA **nucleotide**) occurs within immune cell precursors. This leads to their death through **apoptosis** within six months of birth. Unlike other forms of SCID, ADA-SCID patients can be treated through enzyme replacement therapy. Weekly injections of ADA enzyme (stabilized with polyethylene glycol) hinders toxic deoxyATP buildup, reducing apoptosis and permitting some B- and T-lymphocytes to mature. However, bone marrow transplantation provides better immunity if it succeeds.

JAK3 Deficiency. This accounts for 7 percent of SCID patients. This form of SCID maps to the Janus **kinase** 3 gene on chromosome 19. JAK3 enzyme, a tyrosine kinase, is part of the intracellular signaling pathway (JAK-STAT)

autosomal describes a chromosome other than the X and Y sex-determining chromosomes

cytokine immune system signaling molecule

nucleotide the building block of RNA or DNA

apoptosis programmed cell death

kinase an enzyme that adds a phosphate group to another molecule, usually a protein

Baby David ("bubble boy") Vetter plays in his plastic-enclosed environment. This protection helped David to survive, since he was born with severe combined immune deficiency syndrome.

that activates the genes for T-lymphocyte differentiation. To do so, it must bind to the SCID-X1 gene product.

Interleukin-7 receptor α Chain Deficiency. This is responsible for SCID in a small number of patients. This receptor is part of a larger complex that includes the SCID-X1, JAK3 proteins, and four other interleukin receptors (IL-2, IL-5, IL-9 and IL-15). Thus, at least three of the known SCID genes have been found to disrupt the same receptor complex involved in immune system development.

RAG1 and 2 Deficiencies. These can also lead to SCID. RAG1 and RAG2 are found side-by-side on chromosome 6. Mutations in either can result in SCID. They are involved in the genetic rearrangement of both the T- and B-lymphocyte receptor genes during differentiation that gives rise to the ability of the immune system to recognize foreign agents. Mutations in RAG1 or 2 also account for 50 percent of those patients with an unusual **autoimmunity** form of SCID called Omenn syndrome.

autoimmunity immune reaction to the body's own tissues

The genes involved in two additional rare forms of SCID, reticular dysgenesis and cartilage-hair hypoplasia, remain unknown. In addition, the gene(s) for over 30 percent of those patients diagnosed with SCID also remain unidentified, even though these patients seem to display the same **phenotypes** as those patients whose genetic defect has been identified.

phenotypes observable characteristics of an organism

Gene Therapy

Understanding the genes responsible for SCID has also led to an increased understanding of the genes involved in the overall development of the immune system. It has also helped to unravel the complicated signaling pathways between the cells of the immune system that control and define the immune response itself. There has also been another major benefit. The study of SCID has, in the past, aided in developing an effective program for bone marrow transplantation.

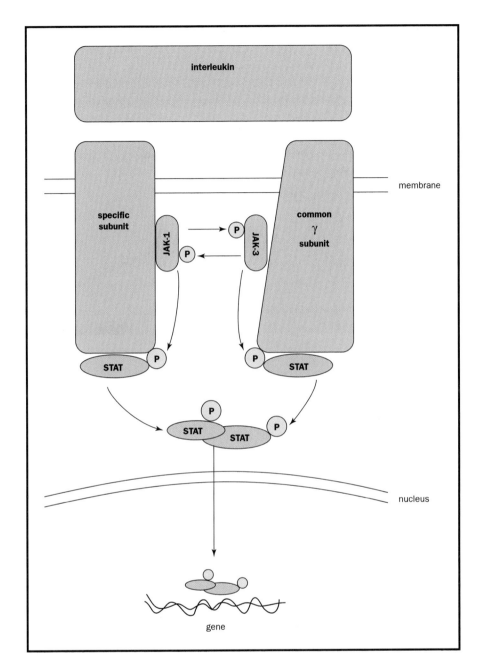

Interleukin binds to its two-part receptor, triggering JAK-1 and JAK-3 to add phosphates to each other and to STAT (signal transducers and activators of transcription). This allows the STATs to combine and enter the nucleus, where they promote gene expression leading to immune system development. SCID can arise from mutations that inactivate either the γ subunit or JAK-3 (shaded structures).

In 1990 the first gene therapy was attempted in two ADA-SCID patients. ADA-SCID was specifically chosen for the first attempted gene therapy for several reasons: bone marrow transplantation had indicated that replacing the defective gene was possible; mouse SCID models had been effectively treated through gene therapy; and ADA-SCID could be treated with some success with enzyme-replacement therapy, enhancing the opportunity for successful gene therapy.

Unfortunately, the majority of ADA-SCID patients, as well as the 3,000 patients enrolled in gene therapy trials for a broad array of other diseases since the first ADA-SCID gene therapy, have not been significantly helped. One exception is the very first gene therapy patient, four-year-old Ashanti de Silva who, ten years later, had a normal lifestyle, with a level of 20 to 25

percent normal T-lymphocytes. Learning from these gene therapy experiences, French researchers modified the procedure. In 2000 they reported successful gene therapy for two infants with SCID-X1. The patients had left the hospital and its protective isolation after a three-month stay. Ten months after gene therapy, they remained healthy, with normal levels of B- and T-lymphocytes and natural killer cells. Thus SCID continues to define the current state of the art for gene therapy, exposing its limitations while simultaneously pointing to its eventual success. SEE ALSO EMBRYONIC STEM CELLS; GENE THERAPY; HIV; IMMUNE SYSTEM GENETICS; RODENT MODELS; SIGNAL TRANSDUCTION.

Diane C. Rein

Bibliography

Fischer, Alain. "Primary Immunodeficiency Diseases: An Experimental Model for Molecular Medicine." *Lancet* 357, no. 9271 (2001): 1863–1869.

Fischer, Alain, et al. "Gene Therapy for Human Severe Combined Immunodeficiencies." *Immunity* 15 (2001): 1–4.

Leonard, Warren J. "X-linked Severe Combined Immunodeficiency: From Molecular Cause to Gene Therapy within Seven Years." *Molecular Medicine Today* 6, no. 10 (2000): 403–407.

Wheelwright, J. "Body, Cure Thyself." *Discover* 23, no. 2 (2002): 62–68.

Sex Determination

Sex determination refers to the mechanisms employed by organisms to produce offspring that are of two different sexes. First we present an overview of the sex determination mechanisms used by mammals. Then we discuss the great variety of mechanisms used by animals other than mammals.

Mammalian Mechanisms

A developing mammalian embryo's gender is determined by two sequential processes known as primary and secondary sex determination.

Primary Sex Determination. Early in an embryo's development (four weeks after fertilization, in humans), two groups of cells become organized into the gonad rudiment that will eventually develop into either the ovaries or testicles. These **gonads** will eventually be the source of **gametes** in the adult. However, at this early stage they are unstructured organs that lack sex-specific features but have the potential to develop into gonads.

The first visible indication of sex-specific development, occurring in week seven in humans, is in males, with the gonads restructuring into two distinct compartments: the testicular cords and the interstitial region. In females, the gonads appear to lack distinct structures until later in development. Primary sex determination, including the differentiation of an embryo's gonads, is dependent on genetic factors associated with the embryo's sex chromosomes.

Secondary Sex Determination. Secondary sex determination involves the development of additional sex-specific characteristics, such as the genitalia. This secondary pathway is controlled by sex-specific hormones that are

gonads testes or ovaries

gametes reproductive cells, such as sperm or egg

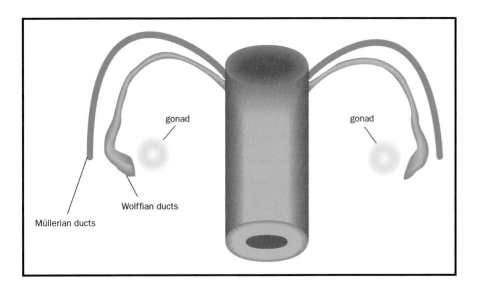

Early in development, every fetus has two sets of primitive ducts. Under the influence of testosterone and the hormone AMH, the Wolffian ducts develop into the male vas deferens and accessory structures. Without testosterone or AMH, the Wolffian ducts degenerate and the Müllerian ducts develop into the female fallopian tubes, uterus, cervix, and upper vagina.

secreted by the differentiated gonad. These hormones influence the sex differentiation of other parts of the body, including two pairs of ducts present in all developing embryos: the Müllerian ducts and the Wolffian ducts.

Testicles produce Müllerian inhibiting substance, a hormone that causes the Müllerian duct to degenerate. They also produce testosterone, which causes the Wolffian duct to develop into the internal male genitalia, such as the seminal vesicles and the vas deferens. Testosterone also promotes the development of the external male genitalia, including the penis, and it reduces the development of the breasts.

In females, where there are no testicles and where there is therefore no Müllerian inhibiting substance, the Müllerian duct differentiates into internal female genitalia: the **fallopian tubes**, uterus, and cervix.

fallopian tubes tubes through which eggs pass to the uterus

Discovering the Testis-Determining Factor. The different effects of the primary and secondary sex determination pathways was demonstrated by embryological transplant experiments carried out by Alfred Jost in the 1940s at the Collège de France in Paris, France. When Jost placed an undifferentiated gonad from a male rabbit next to an undifferentiated gonad inside a female fetus, the gonad from the female developed into an ovary, and the gonad from the male developed into a testicle, as it would have done inside the male. Hence, the sex of the gonad was dependent upon its genotype (XX or XY) and is a result of primary sex determination. The genitalia of these experimental animals revealed the influence of secondary sex determination mechanisms. Under the conflicting signal of the two gonads, the Müllerian duct, which normally would have developed into female genitalia, degenerated partially, and the Wolffian duct, which normally would have degenerated, began to develop into male genitalia.

Jost's experiment indicated that the sex differentiation of a gonad is determined by its sex chromosomes, and that the sexual characteristics of other tissue are determined by the gonads, not by the chromosomes in the tissues themselves. Jost also showed that in the absence of either gonad, the fetus develops as a female. Female development, then, is apparently a "default" pathway that can be overridden to produce a male.

meiosis cell division that forms eggs or sperm

Studies by C. Ford, P. Jacobs, and co-workers in 1959 demonstrated the importance of genes on the X and Y chromosomes as sexual determinants by documenting the sexual phenotypes of humans with abnormal chromosomal constitutions. Errors in **meiosis** can produce sperm or egg cells that have abnormal numbers of sex chromosomes. Upon fertilization, these cells develop into embryos with their own aberrant sex chromosome dosage.

Cells in individuals with Turner's syndrome contain a single copy of the X chromosome and no copies of the Y chromosome. These "XO" individuals develop as females, although their ovaries are nonfunctional. This demonstrates that just a single copy of the X chromosome is sufficient to direct most female sex development.

phenotypic related to the observable characteristics of an organism

The reciprocal condition, "YO," with no X chromosome present, has not been documented in humans, as X chromosomes contain genes necessary for an embryo's survival. Individuals with Klinefelter's syndrome (XXY), however, develop as **phenotypic** males, although they produce no sperm. This indicates that a single copy of the Y chromosome is sufficient to override the female developmental program and promote most male development.

Such observations led to speculation that the Y chromosome contains a "testis-determining factor" necessary to activate development of the male gonads. Several genes on the Y chromosome were suggested as possible testis-determining factors but ultimately rejected. In 1990 Peter Goodfellow and coworkers at the Human Genetics Laboratory in London, England, studied a group of sex-reversed XX males. Such individuals develop as phenotypic males despite being XX individuals.

The researchers discovered that the XX males had a small segment from a Y chromosome incorporated into one of their X chromosomes. This same segment was found to be missing from the Y chromosome of a group of sex-reversed XY individuals, who developed as phenotypic females. The segment acted as a testis-determining factor, as its presence was correlated with the activation of male development. DNA sequencing of the segment identified a gene that was named "*SRY*" from the description "sex-determining region of the Y chromosome."

SRY's Function. Studies in mice have supported *SRY*'s role as a primary determinant of male development. The mouse homolog of *SRY* (Sry) is expressed in developing gonads in males but not females. It is present during but not after testis differentiation. Finally, experiments have been conducted where introducing the *SRY* gene into the genomic DNA of embryonic female (XX) mice causes some of them to develop as males.

transcription messenger RNA formation from a DNA sequence

Despite the clear causal relationship between this gene and male development, the specific mechanisms involved are unclear. The SRY protein is similar to "high-mobility group" proteins, which regulate the **transcription** of other genes. Its structure contains a domain that can bind to specific target DNA sequences. Mutations to this domain, which reduce SRY's ability to bind correctly to DNA, are frequently observed in XY individuals that develop as females.

SRY could conceivably function by activating genes involved in testicle differentiation, by repressing genes involved in ovary development, or by doing both of these things. To discriminate among the possibilities, it is necessary to know more about the next level of genes in the developmental cascade, SRY's targets. Several possible direct targets of SRY have been proposed. Most notable is the *SOX9* gene, which also encodes a high-mobility group protein that is known to promote male differentiation and which is expressed in the developing male gonad immediately after *SRY* is first expressed.

Female Development as the Default. When Jost removed the gonads from embryonic rabbits, the embryos that were XX as well as those that were XY developed as females, though they lacked internal genitalia. This finding emphasized that gonads are critical to secondary sex determination and in the absence of male-specific hormones, female characteristics develop even in an XY individual.

As noted above, individuals with only a single copy of the X chromosome in each cell can survive, and they develop as females. Their ovaries develop normally at first but degenerate around the time of birth, resulting in a sterile adult. Hence a single X chromosome suffices for sex determination, but two copies are needed for ovary maintenance.

One explanation is that the female developmental program is the default, with embryos developing as females unless there are alternative instructions. *SRY* is a major switch gene required for male development, and only a gonad whose cells contain a copy of *SRY* will differentiate into a testicle.

Despite intensive searching, no major switch gene has been identified for the female developmental pathway. One candidate, *DAX1*, was proposed as the main ovarian differentiation gene principally because it was found to be duplicated on the X chromosomes of two XY siblings who developed as females. However, experimentally disrupting the mouse homolog of *DAX1* had no effect on the sex determination, maturation, or fertility of female XX mice.

Instead of having a positive regulatory role, *DAX1* appears to be antagonistic to *SRY*'s function. When *DAX1* is present in two copies, as in the XY sisters, it apparently disrupts *SRY* function sufficiently to prevent initiation of the male developmental program. These observations are consistent with the idea that the female sex determination pathway is the default option.

Nonmammalian Mechanisms

Sex determination and differentiation occur in virtually all complex organisms, but the mechanisms used by various animal classes, and even by various vertebrates, differ significantly. Birds, for instance, lack a clear homolog of *SRY*. In birds it is the female, rather than the male, that has two different sex chromosomes, with males being ZZ, and females ZW. In many reptiles, environmental conditions, rather than genetic factors, are the primary determinant of sex. The temperature at which eggs are incubated determines sex in some lizard, turtle, and alligator species.

In both the fruit fly *Drosophila melanogaster* and the **nematode** *Caenorhabditis elegans*, the primary sex determination mechanisms and the molecular cascades controlling sexual differentiation have been studied

nematode worm of the Nematoda phylum, many of which are parasitic

autosomal describes a chromosome other than the X and Y sex-determining chromosomes

hormones molecules released by one cell to influence another

extensively. Primary sex determination in these animals does not involve the Y chromosome but instead is determined by the ratio of the number of X chromosomes to **autosomal** (nonsex) chromosomes. By examining individuals with unusual numbers of various chromosomes it has been determined that in *Drosophila* those with one or fewer X chromosomes per diploid autosome set develop as males while those with two or more X chromosomes develop as females. Individuals with intermediate ratios such as those with two X chromosomes and a triploid set of autosomes develop as intersexes with both male and female characteristics. Although this ratio serves as the primary determinant of sex in both of these organisms, the specific gene products that influence this ratio assessment are different, demonstrating that different molecular mechanisms can be used for a similar purpose.

There is also significant variability in the strategies by which the outcome of sex determination is communicated to the various tissues that undergo sexual differentiation. In humans and most other mammals, the presence or absence of SRY protein in cells of the gonad specifies their sexual differentiation and which hormones are secreted by the gonad to direct the sexual differentiation of most other cells in the individual.

In *Drosophila* **hormones** have little effect on sexual differentiation. Instead, with only a few exceptions, each cell decides its sex independently of other cells and tissues. This "cell autonomous" mechanism is demonstrated in experimentally produced mosaic organisms called gynandromorphs ("male-female forms") in which some cells are XX (female) and others XO (male). Such individuals develop into adults with a mix of male and female cell types that match each cell's genotype. The lack of evolutionary conservation of sex-determining mechanisms among animals is particularly interesting because of the similarities that exist in other major switch genes for basic developmental processes. SEE ALSO ANDROGEN INSENSITIVITY SYNDROME; GENE EXPRESSION: OVERVIEW OF CONTROL; NONDISJUNCTION; TRANSCRIPTION FACTORS; X CHROMOSOME; Y CHROMOSOME.

Jeffrey T. Villinski and William Mattox

Bibliography

Berta, Philippe, et al. "Genetic Evidence Equating *SRY* and the Testis-Determining Factor." *Nature* 348 (1990): 448–450.

Cline, Thomas W., and Barbara J. Meyer. "Vive la Difference: Males vs. Females in Flies vs. Worms." *Annual Reviews of Genetics* 30 (1996): 637–702.

Gilbert, Scott F. *Developmental Biology*, 6th ed. Sunderland, MA: Sinauer Associates, 2000.

Hodgkin, Jonathan. "Genetic Sex Determination Mechanisms and Evolution." *BioEssays* 4 (1992): 253–261.

Sinclair, Andrew H., et al. "A Gene from the Human Sex-Determining Region Encodes a Protein with Homology to a Conserved DNA-Binding Motif." *Nature* 346 (1990): 240–224.

Zarkower, David. "Establishing Sexual Dimorphism: Conservation amidst Diversity?" *Nature Reviews Genetics* 3 (2001): 175–185.

Sex-Linked Inheritance *See Inheritance Patterns*

Sexual Orientation

The biological basis of sexual orientation (heterosexuality, homosexuality, or bisexuality) has long been a topic of controversy in both science and society. A growing body of research supports the view that genetics and the environment work together to determine sexual orientation. Some issues remain unclear. First, how much of sexual orientation is genetic and how much is shaped by environmental influences, including family, society, and culture? Second, is sexual orientation a fixed trait, or is it subject to environmental influence and changeable over time? Two types of genetic studies, classical family/twin/adoption studies and biological/molecular studies, support multiple genetic and environmental determinants in male and female sexual orientation.

Twin and Family Studies

Twin studies are a classic tool for examining the role of genes. Twins brought up together share a similar environment. **Monozygotic** twins share all their genes, while **dizygotic** twins share only half their genes. Early twin studies by Franz Kallmann in 1952 and Leonard Heston in 1968 reported that if one monozygotic twin was homosexual, there was a greater chance the other twin would be homosexual. The likelihood of this was greater than for dizygotic twins. These studies were potentially biased. They recruited homosexual subjects and had relatively small sample sizes. Recent twin studies have examined all twins in a community without regard to sexual orientation, providing large, less biased sample sizes. In 2000 Kenneth Kendler and colleagues evaluated genetic and environmental factors in a large U.S. sample of twin and nontwin sibling pairs. Sexual orientation was classified as heterosexual or nonheterosexual (bisexual or homosexual) and was determined by a single item on a self-report questionnaire. There was a greater chance for both monozygotic twins to be nonheterosexual than for dizygotic twins or sibling pairs. Results suggested that sexual orientation was greatly influenced by genetic factors, but family environment might also play a role. One problem with this study is that a single item was used to assess the complexity of sexual orientation.

Katherine Kirk's study in 2000 involved a community sample of almost 5,000 adult Australian twins who answered an anonymous questionnaire on sexual behavior and attitudes. Multiple measures of sexual orientation (behaviors, attitudes, feelings) provided stronger evidence for additive genetic influences on sexual orientation. **Heritability estimates** of homosexuality in this sample were 50 to 60 percent in females and 30 percent in males. In 1999 J. Michael Bailey found that if a man was homosexual, the percentage of his siblings who were homosexual or bisexual was 7 to 10 percent for brothers and 3 to 4 percent for sisters, higher than would be due to chance.

Some family studies have reported more homosexuals had homosexual maternal relatives but not paternal relatives. This might support a genetic factor on the X chromosome and/or environmental influences. Other, similar studies did not find this. Thus, evidence exists for both genetic and environmental determinants of sexual orientation which may be different for men and women.

monozygotic genetically identical

dizygotic fraternal or nonidentical

heritability estimates how much of what is observed is due to genetic factors

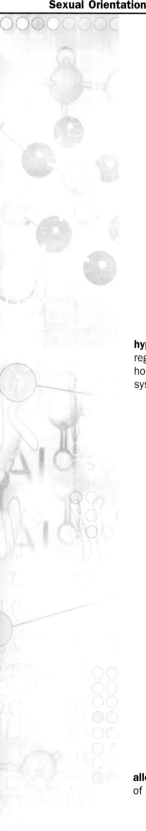

A 2000 study examined whether sexual orientation is fixed or changes with time through environmental influence or the effects of aging. J. Michael Bailey recruited a community sample of twins from the Australian Twin Registry and assessed sexual orientation, childhood gender nonconformity (atypical gender behavior), and continuous gender identity (an individual's self-identification as "male" or "female"). Familial factors were important for all traits, but less successful in distinguishing genetic from shared environmental influences. Only childhood gender nonconformity was significantly heritable for both men and women. Statistical tests suggested that causal factors differed between men and women, and for women provided significant evidence for the importance of genetics factors.

Birth-order studies found homosexual males were not usually first born, having older siblings. Extremely feminine homosexual men had a higher than expected proportion of brothers, not an equal numbers of brothers and sisters.

Biological and Genetic Linkage Studies

hypothalamus brain region that coordinates hormone and nervous systems

Biological studies looking at the **hypothalamus** have found differences between homosexual and heterosexual men and women. Some researchers found differences in parts of the hypothalamus, while others did not. What these findings mean is not clear because they were inconsistent.

In 1995 William Turner examined the ratio of males to females among relatives of the mothers of male homosexuals. He reported that the sex ratio was not normal in maternal relatives. The normal ratio, for relatives of heterosexual males, was an even split: 50 percent male relatives, 50 percent female relatives. The number of male relatives of homosexual males, on the other hand, was significantly lower than the number of female relatives. Also, 65 percent of the mothers of homosexuals had no live-born brothers, or else they had only one live-born brother. On the paternal side, however, the number of male and female relatives of male homosexuals was same as that found for heterosexuals, and the sex ratio of relatives on both the maternal and paternal side for female homosexuals was the same as for heterosexuals. These findings would support genetic factors on the X chromosome, which males inherit from their mothers, as a factor that may cause fetal or neonatal loss of males.

alleles particular forms of genes

Molecular studies found a linkage between male homosexuality and the X chromosome. Dean Hamer and colleagues in 1993 and Nan Hu and colleagues in 1995 conducted DNA linkage analyses in U.S. families with two homosexual brothers. There was significant linkage at Xq28 for 64 percent of homosexual male siblings but not for homosexual females (Xq28 is band 28 of the long arm of the X chromosome). George Rice and colleagues in 1999 examined four **alleles** at Xq28 in fifty-two Canadian male homosexual siblings but did not find any such linkage. This could represent genetic variation, diagnostic differences, and/or different methods of data analysis.

In summary, family, biological, and molecular data support multiple genetic and environmental bases for sexual orientation, and evidence exists for childhood gender nonconformity. SEE ALSO COMPLEX TRAITS; GENE AND THE ENVIRONMENT; PUBLIC HEALTH, GENETIC TECHNIQUES IN; NOMENCLATURE; SEX DETERMINATION; STATISTICS; TWINS; X CHROMOSOME; Y CHROMOSOME.

Harry Wright and Ruth Abramson

Bibliography

Bailey, J. Michael., M. P. Dunne, and N. G. Martin. "Genetic and Environmental Influences on Sexual Orientation and Its Correlates in an Australian Twin Sample." *Journal of Personal and Social Psychology* 78 (2000): 524–536.

Friedman, R. C., and J. I. Downey. "Homosexuality." *New England Journal of Medicine* 331 (1994): 923–930.

Hamer, Dean H., et al. "A Linkage between DNA Markers on the X Chromosome and Male Sexual Orientation." *Science* 261 (1993): 321–327.

Kendler, Kennneth S., et al. "Sexual Orientation in a U.S. National Sample of Twin and Nontwin Sibling Pairs." *American Journal of Psychiatry* 157 (2000): 1843–1846.

Sickle Cell Disease *See Hemoglobinopathies*

Signal Transduction

To survive, an organism must constantly adjust its internal state to changes in the environment. To track environmental changes, the organism must receive signals. These may be in the form of chemicals, such as **hormones** or nutrients, or may take another form, such as light, heat, or sound. A signal itself rarely causes a simple, direct chemical change inside the cell. Instead, the signal sets off a chain of events that may involve several or even dozens of steps. The signal is thereby transduced, or changed in form. Signal transduction refers to the entire set of pathways and interactions by which environmental signals are received and responded to by single cells.

hormones molecules released by one cell to influence another

Signal transduction systems are especially important in multicellular organisms, because of the need to coordinate the activities of hundreds to trillions of cells. Multicellular organisms have developed a variety of mechanisms allowing very efficient and controlled cell-to-cell communication. Though we take it for granted, it is actually astonishing that our skin, for example, continues to grow at the right rate to replace the continuous loss of its surface every day of our lives. This tight regulation is found in every tissue of our body all of the time, and when this fine control breaks down, cancer may be the result. Clearly the molecular mechanisms behind this astounding level of control must be powerful, versatile, and sophisticated.

Signals, Receptors, and Cascades

The signals that cells use to communicate with one another are often small amino acid chains, called **peptides**. Depending on the cell type that releases them and the effect they have on the target cell, they may be called hormones, growth factors, neuropeptides, **neurotransmitters**, or **cytokines**. Other small molecules can also be signals, such as amino acids and steroids such as testosterone. External signals such as odorants and tastes can be carried to us in the atmosphere or in the fluids of our food and drinks. Stretch, pressure, and other mechanical effects as well as heat, pain, and light can also act as signals.

peptides amino acid chain

neurotransmitters molecules released by one neuron to stimulate or inhibit a neuron or other cell

cytokines immune system signaling molecules

Given the huge variety of signals to which a cell is exposed, how does it know which to respond to? The answer is that signals are received by protein receptors made by the cell, and a cell is sensitive only to those signals for which it has made receptors. For instance, every cell in the body is

Signal transduction converts an environmental signal into a cellular response. A cascade allows rapid response, amplification and control.

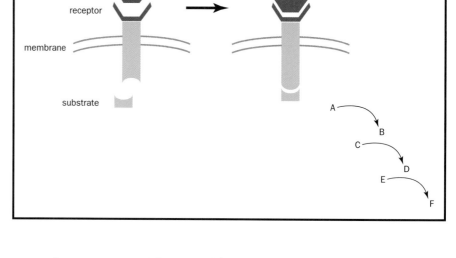

exposed to estrogen circulating in the blood, but only a subset of them make estrogen receptors, and are therefore sensitive to its influence.

Chemical signals such as hormones bind to their receptors, usually at the surface of the cell (the plasma membrane), but sometimes within the cell. This causes a conformation (shape) change in the receptor. The conformation change typically alters the ability of the receptor to bind to another molecule in the cell, modifying that molecule's conformation, or triggering other actions.

This sequence of events triggered by the signal-receptor interaction is called a **transduction** cascade. A transduction cascade involves a network of **enzymes** that act on one another in specific ways to ultimately generate precise and appropriate responses. These responses may include alterations in cell motility or division, induction of the expression of specific genes, and the regulation of **apoptosis**. The molecular details of several such cascades are known, but many more undoubtedly remain to be discovered.

The value of this complex cascade of events is severalfold. First, the network of interactions provides many levels of control, so that the cell can tailor the magnitude and timing of its response very finely. Second, the many levels of interaction allow amplification of the original signal to quickly produce a strong response to a small stimulus. For example, there may be only a few hundred copies of a specific receptor on the surface of a typical cell. Activation of even a small percentage of them, acting through these amplifying enzyme cascades, can result in activation of millions of downstream target molecules. This explains how even very small amounts of signals such as growth factor can have such profound effects on appropriately receptive cells.

The Importance of Phosphorylation and Dephosphorylation

After a signal is received, signal transduction involves altering the behavior of proteins in the cascade, in effect turning them on or off like a switch. Adding or removing phosphates is a fundamental mechanism for altering the shape, and therefore the behavior, of a protein. **Phosphorylation** may open up an enzyme's active site, allowing it to perform chemical reactions,

transduction conversion of a signal of one type into another type

enzymes proteins that control a reaction in a cell

apoptosis programmed cell death

phosphorylation addition of the phosphate group PO_4^{3-}

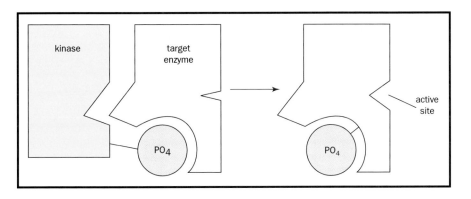

Phosphorylation may open the active site of an enzyme.

or it may frequently generate a binding site allowing a specific interaction (may make a bulge in one side preventing the protein from fitting together) with a molecular partner.

Enzymes that add phosphate groups to other molecules are called **kinases**, and the molecules the enzymes act on are called substrates. Protein kinases are a family of enzymes that use **ATP** to add phosphate groups on to other proteins, thereby altering the properties of these substrate proteins. Protein kinases themselves are frequently turned on or off by phosphorylation performed by other protein kinases; thus a kinase can be both enzyme and substrate.

Protein kinases can be broadly divided into two groups based on the amino acids to which they add phosphate groups. The serine/threonine kinases (ser/thr) are found in all **eukaryotic** cells and tend to be involved in regulation of metabolic and cytoskeletal activity. The tyrosine (tyr) kinases are found in all animals but not in yeast, protozoa, or plants, and appear to have evolved specifically to deal with the complex challenges of signaling in animals.

Molecular switches are useful only if they can also be flipped back to their original state. This is achieved by specific protein phosphatases, which can remove phosphate groups from kinase substrates.

Signal Transduction: The RTK Pathway

How does the cell "know" when a particular receptor molecule in the membrane is occupied, and how is that information chemically translated into actions within the cell? Let us examine the signaling pathway for the receptor tyrosine kinases (RTK). The RTKs are a very powerful and important family of signaling molecules and include receptors for potent growth factors and such hormones as insulin, epidermal growth factor, and nerve growth factor.

In this system, the extracellular "**ligand**" (growth factor or hormone) must crosslink two receptor molecules in order to begin the transduction cascade. The interaction of the two intracellular domains of the receptors then initiates a signaling response.

The simplest RTKs have three parts: a ligand binding site outside the cell, a single membrane-spanning domain, and a tyrosine kinase domain inside the cell. The ligand is typically a diffusible peptide or small protein

kinases enzymes that add a phosphate group to another molecule, usually a protein

ATP adenosine triphosphate, a high-energy compound used to power cell processes

eukaryotic describing an organism that has cells containing nuclei

ligand a molecule that binds to a receptor or other molecule

In the GPCR pathway, an external signal causes an internal, multi-part G protein to exchange GDP for GTP. It then activates adenyl cyclase and ultimately triggers gene transcription.

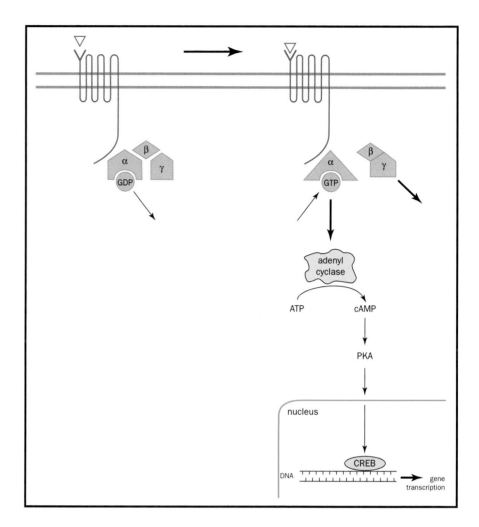

produced elsewhere in the organism, and in this case is the specific growth factor recognized by the RTK. In the absence of its specific growth factor, this receptor remains unbound to a second receptor, and is inactive. Growth factor, when it arrives, binds to two receptors, cross-linking them. This causes the two tyrosine kinase domains to come into contact with one another. Each kinase now has a substrate, formed by the other receptor, and so each phosphorylates the other on multiple tyrosine sites.

The receptors may now bind with one or more other proteins (called SH2 proteins) that specifically recognize their phosphorylated tyrosines. Many of these are enzymes, while others are adapter molecules that in turn attract and bind other enzymes. Often these enzymes are inactive until they join the receptor complex, because their substrates are found only in the membrane. The products of these enzymes may act on yet other molecules, thus continuing the signaling cascade, or they may be used in cell metabolism for growth or other responses.

Signal Transduction: The GPCR Pathway and G Proteins

The G protein coupled receptors (GPCR) illustrate another way to receive and interpret a signal, in which the ligand binds to a single transmembrane protein, causing a conformation change and activating "G proteins" inside the cell. G proteins are so named because they have a binding site for guanine nucleotide, either a diphosphate (GDP) or a triphosphate (GTP). Like ATP, GTP is the high-energy form of the nucleotide, while GDP is the low-energy form.

The GPCR family of proteins has at least 300 members in humans. They are specific receptors for numerous neurotransmitters, hormones, peptides and other substances. A large subgroup of these are odorant receptors directly responsible for our senses of taste and smell.

Surprisingly, a further group of GPCRs is responsible for our sense of vision. Opsin molecules, including rhodopsin, are actually GPCRs. Instead of a ligand binding site, rhodopsin has a molecule of retinal bound in the same relative position. Light alters the conformation of the retinal, and rhodopsin responds to this by altering its protein conformation. This causes G protein activation in a similar manner to the other GPCRs.

Each GPCR is associated with a particular type of G protein inside the cell, and there are dozens of different G proteins known. G proteins are generally inactive in the GDP-bound form. They are activated when GDP departs and is replaced by GTP. GTP is found in excess in cells, and so GDP departure is followed rapidly by GTP binding. This causes a profound conformational change, allowing the G protein to interact with and influence numerous target molecules.

Each GPCR associated G protein consists of three parts, the α, β, and γ subunits. The α subunit binds either GDP or GTP. The β and γ subunits are always found together in a complex called $G\beta\gamma$. In unstimulated cells the whole three-part complex is found in the **plasma membrane**, with GDP in the binding site.

plasma membrane
outer membrane of the cell

Binding of ligand to the GPCR causes a conformational change that is transmitted to the cytoplasmic region of the receptor, which interacts with the G protein to dissociate the GDP. This results in GTP binding, which alters the structure of the α subunit, freeing it from $G\beta\gamma$. Both parts, the $G\alpha$(GTP) and the $G\beta\gamma$, now diffuse away from the receptor and separately interact with and influence many other molecules in the cell.

Second Messengers

One target for certain $G\alpha$(GTP) types is the enzyme adenyl cyclase. This enzyme uses ATP to generate cyclic AMP (cAMP). The cAMP molecule is a "second messenger," one of a family of small diffusible substances that powerfully induce cytoplasmic responses.

Cyclic AMP exerts much of its effects by activating the cAMP-dependent protein kinase (PKA), a ser/thr kinase that can phosphorylate and influence many cellular proteins. For example, PKA phosphorylates CREB (cyclic AMP response element binding protein), which is found attached to the promoters of many genes. Phosphorylation of CREB by PKA can thus regulate the expression of these genes.

Another $G\alpha$(GTP) target is an enzyme, phospholipase C, which cleaves a membrane lipid called PIP-2. This produces two products: DAG and IP-3. DAG stays in the membrane and binds all members of the ser/thr protein kinase C (PKC) family of enzymes, which may then become activated

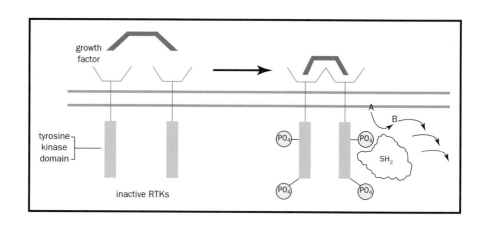

Activated RTK binds SH$_2$ proteins, which can then act as their substrates (A) to form products (B).

endoplasmic reticulum network of membranes within the cell

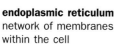

and phosphorylate and regulate a host of metabolic and structural enzymes. IP-3, another second messenger, rapidly diffuses to IP-3 receptors in the **endoplasmic reticulum** membrane. When IP-3 binds, it opens channels in the membrane, releasing stored calcium into the cytoplasm. Calcium is normally kept at very low levels in the cytoplasm, and even small increases cause numerous major effects, so that calcium is also regarded as a powerful second messenger. These effects include the activation of various calcium binding proteins such as calmodulin and its many relatives. The calcium-bound versions of these proteins regulate many other enzymes. For example, calcium activates all members of a large and important family of ser/thr kinases called calcium/calmodulin dependent protein kinases (CAM kinases), which themselves regulate the activity of numerous important substrate molecules.

Just as protein kinases need to be turned off, so too do G proteins. This occurs when the GTP on the Gα is cleaved to generate GDP, thus favoring the reformation of the three-part inactive protein.

Interacting Pathways, Defective Signaling, and Treatments for Disease

The GPCR and RTK pathways do not necessarily remain separate, either from each other or from other signaling pathways. One of the most important pathways is directly downstream of RTKs and also many GPCRs, and is called the mitogen activated protein (MAP) kinase cascade. MAP kinase is an abundant ser/thr kinase that, when activated, phosphorylates and powerfully affects the activity of a large number of cytoskeletal, signaling, and nuclear proteins, including an important family of **transcription factors**, thus directly influencing gene expression.

transcription factors proteins that increase the rate of gene transcription

apoptosis programmed cell death

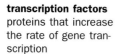

Both pathways also help regulate a particularly important process called **apoptosis**. Many cells need specific growth factors to stay alive; the growth factors trigger pathways involving various ser/thr protein kinases that ultimately inactivate molecules that would otherwise promote apoptosis.

The entire signal transduction system normally works astonishingly well, but serious problems can occur. Cancer is unregulated cell growth and occurs when the machinery tightly regulating cell growth breaks down. It is often easy to see how this has occurred. Mutations in growth factor receptors, G proteins, MAP kinases, and other molecules frequently contribute to cancer, and generally result in these molecules losing their normal switching

function, staying in the activated form and therefore inappropriately stimulating these important enzyme cascades.

The complexity of the signaling system makes for challenging research, but once understood it holds the promise for better treatments for cancer and other diseases. This is because each step in each pathway provides one or more targets for drugs. Designing a drug that could quiet the excess signaling caused by defective MAP kinase, for example, might provide a promising cancer treatment.

The examples given thus far provide only an outline of how signal transduction cascades work and an overview of a few of the most important enzymes. The actual process is much more complex, and there is much about the process that remains mysterious. Perhaps the biggest mystery is how the cell makes sense of all of the input from different growth factors, hormones, extracellular substrates, and so on to produce an appropriate response. The solution to this problem will result from a complete understanding and computer modeling of the biochemical and kinetic properties of the components of all these signaling cascades. SEE ALSO APOPTOSIS; CELL, EUKARYOTIC; CELL CYCLE; GENE; GENE EXPRESSION: OVERVIEW OF CONTROL; PHARMACOGENETICS AND PHARMACOGENOMICS.

Gerry Shaw

Bibliography

Alberts, Bruce., et al. *Molecular Biology of the Cell,* 4th ed. New York: Garland Science, 2002.

Scott, John D., and Tony Pawson. "Cell Communication: The Inside Story." *Scientific American* (June 2000).

Special Issue on Mapping Cellular Signaling. *Science* 296, no. 5573 (May 31, 2002).

Speciation

Speciation is the process by which new species of organisms arise. Earth is inhabited by millions of different organisms, all of which likely arose from one early life-form that came into existence about 3.5 billion years ago. It is the task of **taxonomists** to decide which out of the multitude of different types of organisms should be considered species. The wide range in the characteristics of individuals within groups makes defining a species more difficult. Indeed, the definition of species itself is open to debate.

taxonomists scientists who identify and classify organisms

Concepts of Species

In the broadest sense, a species can be defined as a group of individuals that is "distinct" from another group of individuals. Several different views have been put forward about what constitutes an appropriate level of difference. Principal among these views are the biological-species concept and the morphological-species concept.

The biological-species concept delimits species based on breeding. Members of a single species are those that interbreed to produce fertile offspring or have the potential to do so. The morphological-species concept (from the ancient Greek root "morphos," meaning form) is based on classifying species by a difference in their form or function. According to this

concept, members of the same species share similar characteristics. Species that are designated by this criteria are known as a morphological species.

Organisms within a species do not necessarily look identical. For example, the domestic dog is considered to be one species, even though there is a huge range in size and appearance among the different breeds. For naturally occurring populations of organisms that we are much less familiar with, it is much more difficult to recognize the significance of any character differences observed. Therefore deciding what characteristics should be used as criteria to designate a species can be difficult.

Speciation Mechanisms: Natural Selection and Genetic Drift

Before the development of the modern theory of evolution, a widely held idea regarding the diversity of life was the "typological" or "essentialist" view. This view held that a species at its core had an unchanging perfect "type" and that any variations on this perfect type were imperfections due to environmental conditions. Charles Darwin (1809–1882) and Alfred Russel Wallace (1823–1913) independently developed the theory of evolution by natural selection, now commonly known as Darwinian evolution.

The theory of Darwinian evolution is based on two main ideas. The first is that heritable traits that confer an advantage to the individual that carries them will become more widespread in a population through natural selection because organisms with these favorable traits will produce more offspring. Since different environments favor different traits, Darwin saw that the process of natural selection would, over time, make two originally similar groups become different from one another, ultimately creating two species from one. This led to the second major idea, which is that all species arise from earlier species, therefore sharing a common ancestor.

When so much change occurs between different groups that they are morphologically distinct or no longer able to interbreed, they may be considered different species; this process is known as speciation. A species as a whole can transform over time into a new species (vertical evolution) or split into more separate populations, each of which may develop into new species (adaptive radiation).

genetic drift evolutionary mechanism, involving random change in gene frequencies

Modern population geneticists recognize that natural selection is not the only factor causing genetic change in a population over time. **Genetic drift** is the random change in the genetic composition of a small population over time, due to an unequal genetic contribution by individuals to succeeding generations. It is thought that genetic drift can result in new species, especially in small isolated populations.

Isolating Mechanisms

Whether natural selection and genetic drift lead to new species depends on whether there is restricted gene flow between different groups. Gene flow is the movement of genes between separate populations by migration of individuals. If two populations remain in contact, gene flow will prevent them from becoming separate species (though they may both develop into a new species through vertical evolution).

Gene flow is restricted through geographic effects such as mountain ranges and oceans, leading to geographic isolation. Gene flow can also be prevented by biological factors known as isolating mechanisms. Biological isolating mechanisms include differences in behavior (especially mating behavior), and differences in habitat use, both of which lead to a decrease in mating between individuals from different groups.

When geographic separation plays a role in speciation, this is known as allopatric speciation, from the Greek roots *allo*, meaning separate, and "patric," meaning country. In allopatric speciation, natural selection and genetic drift can act together.

For example, imagine a mud slide that causes a river to back up into a valley, separating a population of rodents into two, one restricted to the shady side of the river, the other to the sunny side. Because coat thickness is a genetically inherited trait, eventually, through natural selection, the population of animals on the cooler side may develop thicker coats. After many generations of separation, the two groups may look quite different and may have evolved different behaviors as well, to allow them to survive better in their respective habitats. Genetic drift may occur especially if either or both populations remain small. Eventually these two populations may be so different as to warrant designation as different species.

It is also possible for new species to form from a single population without any geographic separation. This is known as "ecological" or "sympatric" (from the Greek root *sym*, meaning same) speciation, and it results in ecological differences between morphologically similar species inhabiting the same area. Sympatric speciation can occur in flowering plants in a single generation, due to the formation of a polyploid. **Polyploidy** is the complete duplication of an organism's genome, for example from n chromosomes to $4n$. Even higher multiples of n are possible. This increase in a plant's DNA content makes it reproductively incompatible with other individuals of its former species. SEE ALSO CHROMOSOMAL ABERRATIONS; CONSERVATION BIOLOGY: GENETIC APPROACHES; MUTATION; POPULATION GENETICS; SELECTION.

R. John Nelson

polyploidy presence of multiple copies of the normal chromosome set

Bibliography

Futuyma, Douglas J. *Evolutionary Biology*, 3rd ed. Sunderland, MA: Sinauer Associates, 1998.

Mayr, Ernst. *Evolution and the Diversity of Life: Selected Essays*. Cambridge, MA: Belknap Press, 1976.

Statistical Geneticist

Statistical geneticists are highly trained scientific investigators who are specialists in both statistics and genetics. Training in both statistics and genetics is necessary, as the nature of the work is highly interdisciplinary. Statistical geneticists must be able to understand molecular and clinical genetics, as well as mathematics and statistics, to effectively communicate with scientists from these disciplines.

Typical statistical geneticists spend much of their time working with computers, since much of the statistical analysis of data is now conducted using computer software instead of pencil and paper. Many statistical geneticists are actively engaged in developing new statistical methods for problems that are specific to genetics. Computers are invaluable tools, since much of the initial evaluation of new statistical methods is performed using computer simulations. That is, artificial data are generated using computer algorithms, and the statistical method is evaluated for its ability to identify the specific genetic effects that were simulated. Once a new statistical method is validated using computer simulations it can then be applied to the analysis of real genetic data.

Given the highly interdisciplinary field of statistical genetics, training in multiple different scientific disciplines is necessary. Many statistical geneticists receive four years of undergraduate training in mathematics, statistics, physics, computer science, or some other analytical field of study. This analytical preparation is often necessary for graduate studies in statistical genetics. It is certainly possible, and even desirable in many cases, to receive undergraduate training in biological sciences and then go on to study statistics and genetics in graduate school. There are many different educational paths that a statistical geneticist can take, but the key is interdisciplinary training in an analytical field and genetics. Given the increasing demand for statistical geneticists in the job market, there are many graduate programs that now specialize in statistical genetics.

After finishing four to six years of graduate training, one can follow any of several career paths. Many graduates receive additional training through postdoctoral studies. These studies, known as a postdoc, can last from one to four years and involve working with a research team to gain additional experience in a particular area. For example, if a student had focused primarily on training in mathematics and statistics, it would be possible to do a postdoc to receive the necessary additional training in genetics.

Students who want to target a particular disease area such as cancer for their career might seek out a postdoc that provides experience in cancer research. Students who feel adequately prepared for the job market from their graduate training may be able to go right into a faculty position at a university or to a scientist position in industry without postdoctoral training. Statistical geneticists are in high demand, and there are often more job openings than qualified people to fill them. This will certainly be a growth profession for the next decade.

Statistical genetics is a very exciting professional area because it is so new and there is so much demand. It is a rapidly changing field, and there are many fascinating scientific questions that need to be addressed. Additionally, given the interdisciplinary nature of statistical genetics, there are plenty of opportunities to interact with researchers and clinicians in other fields, such as **epidemiology**, biochemistry, physiology, **pathology**, evolutionary biology, and anthropology. The salary range for statistical geneticists is quite good, since there is so much demand. This is especially true for university medical schools, where starting annual salaries ranged from $70,000 to $80,000 in 2002. Starting salaries for industry positions such as those in pharmaceutical or biotechnology companies were as much as 25 or 50 percent higher.

epidemiology study of incidence and spread of diseases in a population

pathology disease process

A related discipline is genetic epidemiology. Just as statistical genetics requires a combination of training in statistics and genetics, genetic epidemiology requires training in epidemiology and genetics. Since both disciplines require knowledge of statistical methods, there is significant overlap. A primary difference is that statistical geneticists are often more interested in the development and evaluation of new statistical methods, whereas genetic epidemiologists focus more on the application of statistical methods to biomedical research problems. Statistical genetics is an exciting and rewarding career choice for those who have interest and aptitude in the analytical sciences as well as in genetics and biology. SEE ALSO STATISTICS.

Jason H. Moore

Bibliography

Internet Resource

The Committee of Presidents of Statistical Societies (COPSS) Presents Careers in Statistics. American Statistical Association. <http://www.amstat.org/CAREERS/COPSS/>.

Statistics

Statistics is the set of mathematical tools and techniques that are used to analyze data. In genetics, statistical tests are crucial for determining if a particular chromosomal region is likely to contain a disease gene, for instance, or for expressing the certainty with which a treatment can be said to be effective.

Statistics is a relatively new science, with most of the important developments occurring with the last 100 years. Motivation for statistics as a formal scientific discipline came from a need to summarize and draw conclusions from experimental data. For example, Sir Ronald Aylmer Fisher, Karl Pearson, and Sir Francis Galton each made significant contributions to early statistics in response to their need to analyze experimental agricultural and biological data. For example, one of Fisher's interests was whether crop yield could be predicted from meteorological readings. This problem was one of several that motivated Fisher to develop some of the early methods of data analysis. Much of modern statistics can be categorized as exploratory data analysis, point estimation, or hypothesis testing.

The goal of exploratory data analysis is to summarize and visualize data and information in a way that facilitates the identification of trends or interesting patterns that are relevant to the question at hand. A fundamental exploratory data-analysis tool is the histogram, which describes the frequency with which various outcomes occur. Histograms summarize the distribution of the outcomes and facilitate the comparison of outcomes from different experiments. Histograms are usually plotted as bar plots, with the range of outcomes plotted on the x-axis and the frequency of the individual outcome represented by a bar on the y-axis. For instance, one might use a histogram to describe the number of people in a population with each of the different genotypes for the *ApoE* **alleles**, which influence the risk of Alzheimer's disease.

alleles particular forms of genes

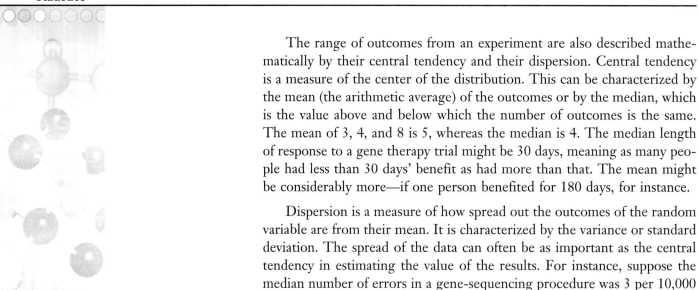

The range of outcomes from an experiment are also described mathematically by their central tendency and their dispersion. Central tendency is a measure of the center of the distribution. This can be characterized by the mean (the arithmetic average) of the outcomes or by the median, which is the value above and below which the number of outcomes is the same. The mean of 3, 4, and 8 is 5, whereas the median is 4. The median length of response to a gene therapy trial might be 30 days, meaning as many people had less than 30 days' benefit as had more than that. The mean might be considerably more—if one person benefited for 180 days, for instance.

Dispersion is a measure of how spread out the outcomes of the random variable are from their mean. It is characterized by the variance or standard deviation. The spread of the data can often be as important as the central tendency in estimating the value of the results. For instance, suppose the median number of errors in a gene-sequencing procedure was 3 per 10,000 bases sequenced. This error rate might be acceptable if the range that was found in 100 trials was between 0 and 5 errors, but it would be unacceptable if the range was between 0 and 150 errors. The occasional large number of errors makes the data from any particular procedure suspect.

Another important concept in statistics is that of populations and samples. The population represents every possible experimental unit that could be measured. For example, every zebra on the continent of Africa might represent a population. If we were interested in the mean genetic diversity of zebras in Africa, it would be nearly impossible to actually analyze the DNA of every single zebra; neither can we sequence the entire DNA of any individual. Therefore we must take a random selection of some smaller number of zebras and some smaller amount of DNA, and then use the mean differences among these zebras to make inferences about the mean diversity in the entire population.

Any summary measure of the data, such as the mean of variance in a subset of the population, is called a sample statistic. The summary measure of the entire group is called a population parameter. Therefore, we use statistics to estimate parameters. Much of statistics is concerned with the accuracy of parameter estimates. This is the statistical science of point estimation.

The final major discipline of statistics is hypothesis testing. All scientific investigations begin with a motivating question. For example, do identical twins have a higher likelihood than fraternal twins of both developing alcoholism ?

From the question, two types of hypotheses are derived. The first is called the null hypothesis. This is generally a theory about the value of one or more population parameters and is the status quo, or what is commonly believed or accepted. In the case of the twins, the null hypothesis might be that the rates of concordance (i.e., both twins are or are not alcoholic) are the same for identical and fraternal twins. The alternate hypothesis is generally what you are trying to show. This might be that identical twins have a higher concordance rate for alcoholism, supporting a genetic basis for this disorder. It is important to note that statistics cannot prove one or the other hypothesis. Rather, statistics provides evidence from the data that supports one hypothesis or the other.

Much of hypothesis testing is concerned with making decisions about the null and alternate hypotheses. You collect the data, estimate the parameter, calculate a test statistic that summarizes the value of the parameter estimate, and then decide whether the value of the test statistic would be expected if the null hypothesis were true or the alternate hypothesis were true. In our case, we collect data on alcoholism in a limited number of twins (which we hope accurately represent the entire twin population) and decide whether the results we obtain better match the null hypothesis (no difference in rates) or the alternate hypothesis (higher rate in identical twins).

Of course, there is always a chance that you have made the wrong decision—that you have interpreted your data incorrectly. In statistics, there are two types of errors that can be made. A type I error is when the conclusion was made in favor of the alternate hypothesis, when the null hypothesis was really true. A type II error refers to the converse situation, where the conclusion was made in favor of the null hypothesis when the alternate hypothesis was really true. Thus a type I error is when you see something that is not there, and a type II error is when you do not see something that is really there. In general, type I errors are thought to be worse than type II errors, since you do not want to spend time and resources following up on a finding that is not true.

How can we decide if we have made the right choice about accepting or rejecting our null hypothesis? These statistical decisions are often made by calculating a probability value, or p-value. P-values for many test statistics are easily calculated using a computer, thanks to the theoretical work of mathematical statisticians such as Jerzy Neyman.

A p-value is simply the probability of observing a test statistic as large or larger than the one observed from your data, if the null hypothesis were really true. It is common in many statistical analyses to accept a type I error rate of one in twenty, or 0.05. This means there is less than a one-in-twenty chance of making a type I error.

To see what this means, let us imagine that our data show that identical twins have a 10 percent greater likelihood of being concordant for alcoholism than fraternal twins. Is this a significant enough difference that we should reject the null hypothesis of no difference between twin types? By examining the number of individuals tested and the variance in the data, we can come up with an estimate of the probability that we could obtain this difference by chance alone, even if the null hypothesis were true. If this probability is less than 0.05—if the likelihood of obtaining this difference by chance is less than one in twenty—then we reject the null hypothesis in favor of the alternate hypothesis.

Prior to carrying out a scientific investigation and a statistical analysis of the resulting data, it is possible to get a feel for your chances of seeing something if it is really there to see. This is referred to as the power of a study and is simply one minus the probability of making a type II error. A commonly accepted power for a study is 80 percent or greater. That is, you would like to know that you have at least an 80 percent chance of seeing something if it is really there. Increasing the size of the random sample from the population is perhaps the best way to improve the power of a study. The closer your sample is to the true population size, the more likely you are to see something if it is really there.

Thus, statistics is a relatively new scientific discipline that uses both mathematics and philosophy for exploratory data analysis, point estimation, and hypothesis testing. The ultimate utility of statistics is for making decisions about hypotheses to make inferences about the answers to scientific questions. SEE ALSO GENE DISCOVERY; GENE THERAPY: ETHICAL ISSUES; STATISTICAL GENETICIST; TWINS.

Jason H. Moore

Bibliography

Gonick, Larry, and Woollcott Smith. *The Cartoon Guide to Statistics.* New York: Harper Collins, 1993.

Jaisingh, Lloyd R. *Statistics for the Utterly Confused.* New York: McGraw-Hill, 2000.

Salsberg, David. *The Lady Tasting Tea: How Statistics Revolutionized Science in the Twentieth Century.* New York: W. H. Freeman, 2001.

Internet Resource

HyperStat Online: An Introductory Statistics Book and Online Tutorial for Help in Statistics Courses. David M. Lane., ed. <http://davidmlane.com/hyperstat/>.

Stem Cells *See Embryonic Stem Cells*

Tay-Sachs Disease

Tay-Sachs disease is a severe genetic disease of the nervous system that is nearly always fatal, usually by three to four years of age. It is caused by mutations in the *HEXA* gene, which codes for a component of the enzyme β-hexosaminidase A or "Hex A." The resulting accumulation of a brain **lipid** called G_{M2} ganglioside produces brain and spinal cord degeneration. It is a rare disease that is found in all populations, but it is particularly prevalent in Ashkenazi Jews of Eastern European origin. There is no treatment, but research aimed at treating the disease by blocking synthesis of the affected molecules has been ongoing since the late 1990s. Carriers can be identified by DNA or **enzyme** tests and prenatal diagnosis is available to at-risk families.

lipid fat or waxlike molecule, insoluble in water

enzyme a protein that controls a reaction in a cell

retina light-sensitive layer at the rear of the eye

History and Disease Description

In 1881 Warren Tay, a British ophthalmologist, observed a "cherry red spot" in the **retina** of a one-year-old child with mental and physical retardation. Later, in 1896 Bernard Sachs, an American neurologist, observed extreme swelling of neurons in autopsy tissue from affected children. He also noted that the disease seemed to run in families of Jewish origin. Both physicians were describing the same disease, but it was not until the 1930s that the material causing the cherry-red spot and neuronal swelling was identified as a ganglioside lipid and the disease could be recognized as an "inborn error of metabolism." The term "ganglioside" was coined because of the high abundance of the brain lipid in normal ganglion cells (a type of brain cell). In the 1960s, the structure of the Tay-Sachs ganglioside was identified and given the name "GM2 ganglioside" (Figure 1).

Gangliosides are glycolipids. The lipid component, called ceramide, sits in the membranes of cells. Attached to it and sticking out into the extra-

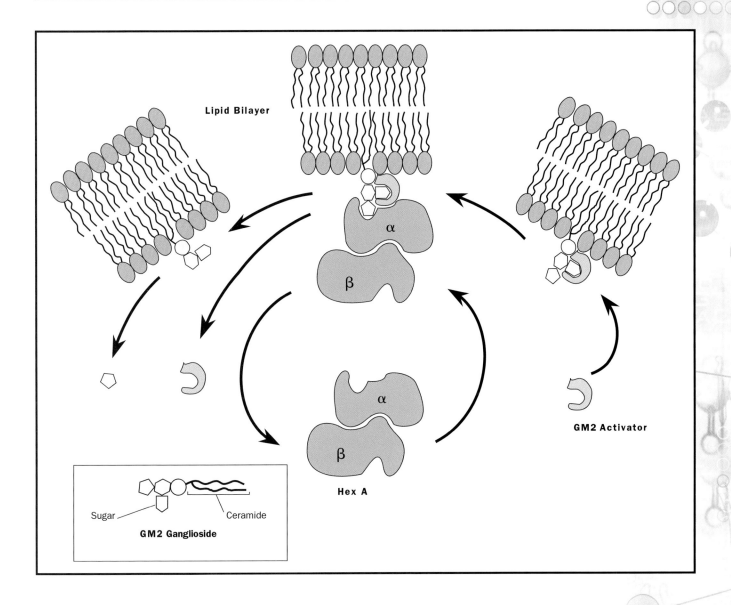

Lipid Bilayer

α

β

α

β

Hex A

Sugar

Ceramide

GM2 Ganglioside

GM2 Activator

cellular space is a linked series of different sugars, the "glyco" portion of glycolipid. The basic function of gangliosides is not well understood, but they appear to have roles in biological processes as diverse as cell-to-cell recognition, differentiation, and in the repair of damaged neurons.

Gangliosides, like most cell components, are broken down and regenerated as part of normal cellular metabolism. The breakdown or "catabolism" of gangliosides occurs in the lysosome, a specialized vesicle that is analogous to the vacuole of plants. In the lysosome a series of acid hydrolases (degradative enzymes) removes each sugar, one at a time, until the ceramide lipid is all that remains. In Tay-Sachs disease, one of the lysosomal hydrolases, Hex A, is defective or completely absent, so the degradative process is blocked before completion. The result is the accumulation of G_{M2} ganglioside, the last molecule before the Hex A block in the catabolic sequence.

Since breakdown is blocked while synthesis continues, the result is a progressive accumulation of GM2 ganglioside and massive swelling of the lysosomes and hence of the neurons containing them. This is the basis of

Figure 1. Molecular basis for Tay-Sachs disease. Absence of or defect in the Hex A protein prevents complete processing of GM2 ganglioside.

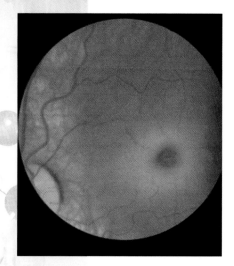

Figure 2. The macular cherry-red spot on the eye is an indicator of Tay-Sachs disease, an autosomal recessive disease that affects the nervous system.

polypeptide chain of amino acids

base pairs two nucleotides (either DNA or RNA) linked by weak bonds

endoplasmic reticulum network of membranes within the cell

Golgi network system in the cell for modifying, sorting, and delivering proteins

the neuron swelling observed by Sachs and the cherry-red spot described by Tay. The cherry-red spot is due to the white appearance of swollen neurons of the retina surrounding the normally red fovea centralis (central depression in retina and site of maximum vision acuity) in the back of the eye (Figure 2).

Newborns with Tay-Sachs disease appear normal at birth. By six months of age, parents begin to notice that their infant is becoming less alert and is less responsive to stimuli. The affected infant soon begins to regress and shows increasing weakness, poor head control, and inability to crawl or sit. The disease continues to progress rapidly through the first years of life, with seizures and increasing paralysis. The child eventually progresses to a completely unresponsive vegetative state. Death is often caused by pneumonia because of the child's weakened state. Some forms of Tay-Sachs disease are much milder with onset of the disease later in childhood or even adulthood. We now know that these forms of the disease are caused by less severe mutations in the *HEXA* gene.

Molecular Biology: Understanding Tay-Sachs Disease

Hex A is composed of two **polypeptide** subunits, one called α and one called β (Figure 1). One other form of the enzyme, Hex B, is composed of two β subunits. In Tay-Sachs disease, it is the α subunit that is mutated so that patients have a defective Hex A, while Hex B remains unaffected. However, Hex B is not active toward G_{M2} ganglioside and can not substitute for Hex A. Some patients have a disease similar to Tay-Sachs, with the absence of both Hex A and Hex B. This condition, now called Sandhoff disease, was first described by Konrad Sandhoff in the 1960s and is due to mutations in the β subunit.

One more protein is involved in the disease. It is called the G_{M2} activator and is essential to the breakdown of G_{M2} ganglioside. The protein forms a complex with the G_{M2} ganglioside, converting the G_{M2} from a hydrophobic, membrane-liking molecule to one that is hydrophilic (water loving) so that it can be successfully hydrolyzed by Hex A in the lysosome. Mutations in the G_{M2} activator gene, called *GM2A*, can also cause a Tay-Sachs–like disease, although it is exceedingly rare.

In sum, mutations in any of three genes can cause the disease: *HEXA*, *HEXB*, or *GM2A*. As a group, patients with any of these diseases are said to have GM2 gangliosidosis. Tay-Sachs disease refers specifically to the most common form of the disease, caused by mutations in the *HEXA* gene.

The *HEXA* gene is one of about 30,000 to 70,000 genes in the human genome. It is of average size at about 35,000 **base pairs** in length and contains fourteen exons. The remainder of the gene is made up of thirteen introns that separate the exons from one another. Extensive research has given us a clear picture how the enzyme is synthesized, processed through the **endoplasmic reticulum** and **Golgi network** of the cell, and sent to the lysosome. This understanding of the cell biology of Hex A has had an important impact on our understanding of mutations in Tay-Sachs disease. Some affect enzyme function, that is, they occur near the "active" site of the enzyme and block its activity, while others affect the biosynthetic processing of the protein. The latter type of mutations may not affect enzyme activity at all. but causes disease because the enzyme fails to reach the lysosome to carry out its biological role.

Mutations and Founder Effect

To date, nearly 100 mutations have been identified in Tay-Sachs disease. Ashkenazi Jews have two common mutations that cause the severe, infantile form of the disease. One, accounting for 80 percent of mutant **alleles** carried in the population, is the loss of four nucleotides in exon 11 of the gene. This causes a "frameshift" in the reading of the genetic code and the inability to generate a complete protein. The second mutation, accounting for about 15 to 18 percent of mutations, is a splice junction mutation, a defect in processing nuclear RNA to form the mature messenger RNA that makes its way to the cytoplasm to direct protein synthesis. When this mutation is present, splicing of intron 12 fails to occur properly, and a functional protein fails to be synthesized. In both cases, the result is the absence of the α subunit and hence Hex A.

alleles particular forms of genes

Ashkenazi Jews and other populations also have mutations that cause amino acid substitutions. One such mutation, accounting for about 3 percent of mutations in Ashkenazi Jews, results in a glycine to serine substitution in exon 7. This mutant α subunit is synthesized and an abnormal Hex A is produced. It is sufficiently active so that patients with this mutation as one of their two mutant alleles have sufficient "residual" Hex A activity to produce a mild, adult form of the disease.

This finding of three predominant mutations causing Ashkenazi Jewish Tay-Sachs disease was unexpected. Most medical scientists thought a single mutation would have acted as a "founder" mutation and over time increased in frequency to the level at which it is found today. It is believed that the first Tay-Sachs mutation may have entered the population about 1000 years ago. It increased in frequency either by random **genetic drift** or possibly through selection for presence of the gene in heterozygous carriers. This latter interpretation is controversial, but it has been suggested that carriers might have been more resistant to tuberculosis than normal individuals so that they had a greater chance of surviving the epidemics of centuries ago, thereby resulting in a steady increase in the frequency of the mutant allele in Ashkenazi Jews. Another group with a founder mutation, a large deletion of the 5′ ("five prime," or front) end of the gene, are French Canadians from the Lac Saint-Jean region of Quebec.

genetic drift evolutionary mechanism, involving random change in gene frequencies

Carrier Testing and Prenatal Diagnosis

One in thirty Ashkenazi Jews is a carrier of one of the Tay-Sachs mutations. This is about ten times the frequency of carriers in non-Jews. Until 1970 it is estimated that about one in 4,000 births among Ashkenazi Jews was of a Tay-Sachs baby. This produced a great desire to develop a carrier and prenatal test shortly after the enzyme defect was identified. Michael Kaback spearheaded a carrier testing program that, by 2000, had tested well over 1 million Ashkenazi Jews, mainly in North America and Israel. This led to a drop in the incidence of Tay-Sachs disease to less than one-tenth of its previous level.

The testing program has been so successful because it is organized through Jewish community groups, with the active participation of geneticists who conduct the tests. During the 1970s and 1980s the test measured the level of Hex A activity in serum or white blood cells. With the identification of the

Ashkenazi mutations, DNA testing came into use. Many geneticists prefer to conduct both types of tests, especially for non-Jews. They find DNA testing useful for its simplicity and exceedingly low error rate, but also recommend enzyme testing to guard against the involvement of a previously undetected mutation that would be missed by the mutation-specific DNA tests.

Future Prospects

In the 1990s laboratories in the United States, Canada, and France developed mouse models of Tay-Sachs disease, Sandhoff disease and G_{M2} activator deficiency. These investigations led to a much better understanding of the brain pathology and progression of the diseases. A significant outcome has been the use of the mouse models to experiment with approaches to therapy. A promising approach is based on partially blocking the synthesis of gangliosides with drugs so that accumulation of G_{M2} ganglioside becomes minimal. In addition to "substrate deprivation," as this blocking action is called, other laboratories are trying gene therapy and drug-based methods for bypassing the Tay-Sachs defect. The combination of carrier testing and prenatal diagnosis to assure the birth of healthy babies, and the more recent prospects for treating affected patients are major advances since the discovery of a cherry-red spot described in the first infant known to have been born with Tay-Sachs disease. SEE ALSO CELL, EUKARYOTIC; DISEASE, GENETICS OF FOUNDER EFFECT; GENE; GENETICS OF DISEASE; HETEROZYGOTE ADVANTAGE; METABOLIC DISEASE; MUTATION; POPULATION SCREENING; PRENATAL DIAGNOSIS; PROTEINS; RNA PROCESSING; RODENT MODELS.

Roy A. Gravel

Bibliography

Gravel, Roy A., et al. "The G_{M2} Gangliosidoses." In *The Metabolic and Molecular Bases of Inherited Diseases*, 8th ed., Charles R. Scriver, Arthur L. Beaudet, William S. Sly and David Valle, eds. New York: McGraw-Hill, 2001.

Internet Resources

"HEXA Locus Database." <http://data.mch.mcgill.ca/gm2-gangliosidoses>.

"Tay-Sachs Disease." National Center for Biotechnology Information, Division of Online Mendelian Inheritance in Man. <http://www.ncbi.nlm.nih.gov:80/entrez/dispomim.cgi?id=272800>.

Technical Writer

A technical writer (sometimes called a technical communicator) designs, writes, edits, and produces documents for scientific, technical, industrial, and government organizations. These documents can include technical reports, specifications, reference manuals, operating instructions, policies and procedures, proposals, presentations, brochures, and Web pages.

Goals and Skills Required

The main aim of a technical writer is to communicate scientific and technical information to other people using easily understandable language. To be a technical writer, a person needs strong language skills, demonstrated by college-level training. A college degree in English, journalism, or communication is preferred, according to the U.S. Department of Labor. The

person also should also have some familiarity with scientific or technical topics. Finally, the person should have experience using word processing and desktop publishing software, graphics programs, and Web publishing tools.

Specially trained people began to be employed as technical writers in the late 1930s. Prior to the 1980s, however, most technical documents still were written by scientists, engineers, and other specialists, many of whom found it difficult to write for nontechnical audiences. With the rapid expansion of science and technology, however, the need increased for people who could both understand complex ideas and convey them effectively to a variety of audiences.

Challenges, Advantages, and Drawbacks

Some challenges need to be considered when thinking about a career as a technical writer. For example, a person may have to invest considerable time and money to acquire the knowledge and skills needed. Also, it can be difficult to gain entry-level experience. Technical writing is typically a sedentary profession that does not involve travel. At the same time, it is a demanding profession that can take time and energy away from other, more creative writing pursuits. Working for a company with an established set of document guidelines can be frustrating, and the profession is sometimes criticized for being dry and unimaginative.

Generally, however, the outlook for technical writers is bright. Technical writing is a job growth area: More jobs are being created than are being filled, particularly in the high technology industry. Once employed, a technical writer works on a wide variety of projects, many of which represent the cutting edge of science and technology. The field is supportive of female professionals; more than half of all technical writers are women. While a majority of technical writers are between the ages of twenty-five and forty-four, about 20 percent are over fifty-five years.

Entering the Profession

In addition, technical writing is a profession that pays well. According to a 2000 salary survey by the Society for Technical Communication, the average salary for a technical writer in the United States. is about $52,000. An entry-level technical writer makes about $37,000, which compares favorably with entry-level positions in other fields. The average salary for a senior-level technical writer with supervisory responsibilities is about $65,000. Salary level also depends on geographic location, level of education, and years of experience in the technical writing field. People interested in seeking employment as technical writers should pursue volunteer and internship opportunities, develop a portfolio of their work to show potential employers, check classified advertisements and company Web sites for job openings, write directly to personnel departments, and/or sign up with a job placement agency that specializes in information technology. SEE ALSO SCIENCE WRITER.

Cindy T. Christen

Bibliography

Society for Technical Communication. "Salary Survey 2000." Supplement to *intercom* 47, no. 8 (2000).

Internet Resources

Conroy, Gary. "Technical What?" Technical Writing 1997. <http://www.techwriting.about.com>.

Kolunovsky, Nina. "Becoming a Technical Writer in Three Easy Steps." Society for Technical Communication 1996. <http://www.stctoronto.org>.

U.S. Department of Labor, Bureau of Labor Statistics. "Writers and Editors, Including Technical Writers." Occupational Outlook Handbook 2000. <http://stats .bls.gov/oco>.

Telomere

Telomeres are structures found at the ends of chromosomes in the cells of **eukaryotes**. Telomeres function by protecting chromosome ends from recombination, fusion to other chromosomes, or degradation by **nucleases**. They permit cells to distinguish between random DNA breaks and chromosome ends. They also play a significant role in determining the number of times that a normal cell can divide. Unicellular forms whose cells have no true nuclei (prokaryotes) possess circular chromosomes that, therefore, have no ends. Thus, prokaryotes can have no telomeres.

eukaryotes organisms with cells possessing a nucleus

nucleases enzymes that cut DNA or RNA

Structure

Telomeres are extensions of the linear, double-stranded DNA molecules of which chromosomes are composed, and are found at each end of both of the chromosomal strands. Thus, one chromosome will have four telomeric tips. In humans, the forty-six chromosomes are tipped with ninety-two telomeric ends.

In most eukaryotic forms, telomeres consist of several thousand repeats of the specific nucleotide sequence TTAGGG and occur in organisms ranging from slime molds to humans. The entire length of repeated telomere sequences is known as the terminal restriction fragment (TRF). Sequences different from TTAGGG are found in more primitive eukaryotic forms, such as the ciliated protozoan *Tetrahymena*, in which Elizabeth Blackburn first characterized the repeated telomere sequence.

polymerases enzyme complexes that synthesize DNA or RNA from individual nucleotides

buffers substances that counteract rapid or wide pH changes in a solution

The **polymerases** that copy the chromosomes of DNA strands are unable to copy completely to the end. This became known as the "end-replication problem" when it was first recognized in the late 1960s. The TRF acts like a **buffer** that protects the information-containing genes, so that the loss of some telomeric nucleotide sequences at each round of DNA replication does not result in the loss of genetic information. The telomeres themselves end in large duplex loops, called T-loops.

A Simple Counting Mechanism

media (bacteria) nutrient source

For the first half of the twentieth century it was believed that cells cultured in laboratory glassware could replicate indefinitely if the correct nutrient **media** and other conditions of growth could be found. Repeated initial failure at culturing indefinitely replicating cells was followed by success in the late 1940s, when the immortal L929 cancer cell population was developed from mouse tissue. Later, other immortal cell populations were found, including the first human cell line, HeLa, derived from a human cervical carcinoma.

It was originally, but erroneously, believed that normal cells also had the potential to divide and function indefinitely in culture, and so it was thought that aging could not be the result of events that occurred within normal cells. Instead, aging was thought to be the result of extracellular events such as radiation or of changes in the extracellular molecules that cement cells to each other.

In 1960, however, it was discovered that no culture conditions exist that will permit normal human cells to divide indefinitely. Rather, cells were found to have a built-in counting mechanism, called the Hayflick Limit, that limits their capacity to replicate. For example, human **fibroblast** cell populations, found in virtually all tissues, will double only about 50 times in culture when derived from fetal tissue. Fibroblast populations from older adults double fewer times, the exact number of doublings depending upon the age of the donor. Leonard Hayflick and P. S. Moorhead also suggested that only abnormal or cancer cells divide indefinitely. They theorized that the limited capacity for normal cells to divide is an expression of aging and that it determines the longevity of the organism.

In support of this theory, it was found that frozen normal fetal cells "remember" the doubling level at which they were frozen and, after thawing, will undergo additional doublings until the total of fifty is reached. These facts suggested to Hayflick that a replication-counting mechanism existed. Hayflick and coresearcher Woodring Wright later found that this mechanism was located in the nucleus of the cell.

The Discovery of Telomeres

The search for the molecular counting mechanism ended when Calvin Harley and Carol Greider discovered that the telomeres of cultured normal human fibroblasts become shorter each time the cells divide. When telomeres reach a specific short length, they signal the cell to stop dividing. Therefore, cellular aging, as marked by telomere shortening, is not based on the passage of time. Instead, telomere loss measures rounds of DNA replication. For this reason, Hayflick has coined the term "replicometer" for this mechanism.

An accumulation of evidence suggests that while telomere attrition explains the loss of replicative capacity in normal cells, the process may not be as simple as first believed. There are several essential DNA-binding proteins (for example, TRF1 and TRF2) associated with telomeres, and the role that they play in capping and uncapping the telomere ends undoubtedly will be found to complicate the oversimplified explanation given above.

Telomerase

Immortal cancer cells escape telomere loss by switching on a gene that expresses an enzyme called telomerase. This unusual enzyme is a **reverse transcriptase** that has an RNA template and a catalytic portion. At each round of DNA replication, telomerase adds onto the existing telomeres the nucleotides that would otherwise have been lost, thus maintaining a constant telomere length. In other words, telomerase acts as an "immortalizing" enzyme. In addition, it has several associated proteins whose roles are still under investigation.

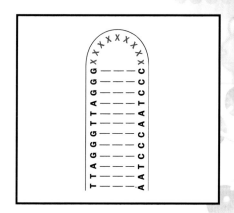

The telomere has several thousand repeated TTAGGG sequences. The loop at the very tip is less well characterized.

fibroblast undifferentiated cell normally giving rise to connective tissue cells

reverse transcriptase enzyme that copies RNA into DNA

somatic nonreproductive; not an egg or sperm

stem cell cell capable of differentiating into multiple other cell types

germ cells cells creating eggs or sperm

fibroblasts undifferentiated cells that normally give rise to connective tissue cells

Using what is called the TRAP assay (telomeric repeat amplification protocol), it has been found that about 90 percent of all human tumors produce telomerase, whereas the only normal adult **somatic** cells that produce telomerase are **stem cell** populations found, for example, in skin, the hematopoietic system, **germ cells**, and gut epithelia. In fact, the presence or absence of telomerase is the most specific property that distinguishes cancer cells from normal cells. This difference is currently under investigation as a diagnostic tool. If a chemical could be found to interfere with telomerase activity in cancer cells, an effective control of this disease might be found. Several candidate substances have been identified and are undergoing extensive studies in animals.

Telomerase is switched on in virtually all human cells at the moment of conception, but as the embryo matures the telomerase becomes repressed in all but the germ cells and stem cell populations. Further, the level of telomerase expressed in stem cells is much less than that expressed in cancer cells. Interestingly, telomerase expression has been found to occur in all the cells of animals that age slowly or not at all. These are animals, such as the American lobster and the rainbow trout, that do not stabilize at a fixed size in adulthood.

On the human genome, an enzyme known as human telomerase reverse transcriptase (hTERT) is found on the most distal gene on chromosome 5p. The transfection (introduction) of hTERT into cultured normal human **fibroblasts** has resulted in telomere elongation, telomerase expression, and the immortalization of these otherwise mortal cells. After several hundred population doublings, the transfected cells exhibit some drift from the diploid number of chromosomes but cancer cell properties do not occur. This experiment proves that telomerase is not a cancer enzyme but an immortalization enzyme. The ability to immortalize normal human cells via hTERT has important potential applications. Some immortalized cells could be cultured in the lab to produce therapeutically useful molecules. Others might be used directly within the body to repair tissue or replace lost or damaged cells. SEE ALSO CHROMOSOME, EUKARYOTIC; DNA POLYMERASES; REPLICATION; REVERSE TRANSCRIPTASE.

Leonard Hayflick

Bibliography

Bodnar, A. G., et al. "Extension of Life Span by Introduction of Telomerase into Normal Human Cells." *Science* 279 (1998): 349–352.

Greider, Carol W. "Telomeres and Senescence: The History, the Experiment, the Future." *Current Biology* 8 (1998): 178–181.

Hayflick, Leonard. *How and Why We Age.* New York: Ballantine Books, 1996.

———. "The Illusion of Cell Immortality." *British Journal of Cancer* 83 (2000): 841–846.

Transcription

Transcription is the process in which genetic information stored in a strand of DNA is copied into a strand of RNA. The sequence of the four bases in DNA, which are adenine (A), cytosine (C), guanine (G), and thymine (T), is preserved in the sequence of the four bases in RNA, which are A, C, G, and uracil (U).

Functions of RNA Transcripts

RNA molecules have various functions in the cell. Many of the functions are associated with translation, in which the genetic code of messenger RNA molecules is used to help the **ribosomes** synthesize a specific protein. In addition, ribosomal RNA is the main component of the ribosome, and transfer RNA does the actual translating from nucleotide sequence into amino acid sequence.

RNA molecules may also function as enzymes. They do so either alone or in association with proteins. RNA molecules associate with proteins, for example, when they serve as components of machinery that helps make other, newly formed RNA molecules functional.

RNA is chemically better suited to carry out certain tasks than is DNA. There are also other reasons RNA, not DNA, is used for these tasks. First, it is desirable to keep DNA available for replication and not tied up with other functions. Second, the small number of DNA molecules in the cell is often insufficient. Creating many identical RNA molecules that are copies of a single segment of DNA provides the necessary numbers. Third, RNA can be differentially degraded when it is no longer needed, providing an important regulatory mechanism that would be unavailable if there were only one type of nucleic acid.

Promoters

Transcription is initiated at regions of DNA called promoters, which are typically 20 to 150 **base pairs** long, depending on the organism. The sequence of bases at a promoter is recognized by RNA polymerase, the enzyme that synthesizes RNA.

The RNA polymerases in bacteria, as well as in viruses in bacteria, are able to recognize particular promoter sequences without the help of any other cellular proteins. However, in eukaryotes and Archaea, other proteins, called initiation factors, recognize the promoter sequence, "recruit" RNA polymerase and other proteins, help the RNA polymerase bind to the DNA, and regulate the enzyme's activity.

RNA polymerase is assembled on promoters in a particular orientation (Figure 1A). This allows RNA synthesis to start at a precise location and proceed in only one direction, "downstream" toward the gene (Figure 1B).

RNA synthesis

RNA, like DNA, is a **polymer** of nucleotides. Each nucleotide consists of a sugar that is attached to a phosphate group and any one of four bases. The RNA polymerase, as it builds the chain of nucleotides, processes only one of the two **complementary** strands of DNA. This DNA strand is referred to as the template strand. The least confusing name for the other DNA strand is "the nontemplate strand."

The bases in the newly synthesized RNA are complementary to the bases in the template DNA strand and, therefore, identical in sequence to the bases in the nontemplate strand, except that the RNA contains U where the nontemplate strand of DNA contains T.

ribosomes protein-RNA complexes at which protein synthesis occurs

base pairs two nucleotides (either DNA or RNA) linked by weak bonds

polymer molecule composed of many similar parts

complementary matching opposite, like hand and glove

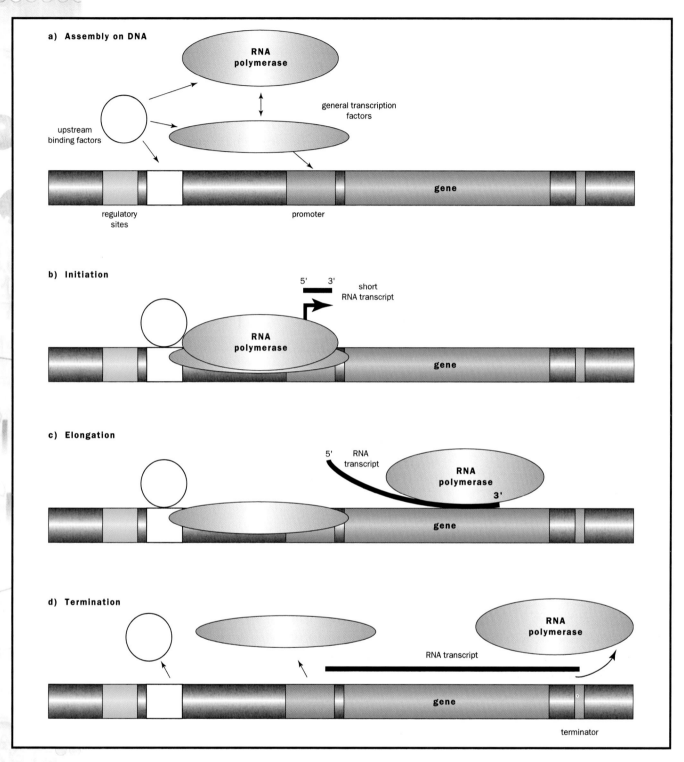

Figure 1. Various steps in the process of transcription. (A) Assembly of the initiation complex. Depending on the conditions RNA polymerase can either bind free general transcription factors and then the promoter, or the general transcription factors that are already promoter-bound themselves. In either case, the upstream binding factors facilitate the interaction of RNA polymerase with the promoter. Note that only one of the upstream regulatory sites is depicted as having a regulatory factor bound. (B) Initiation of RNA synthesis. At precise locations as determined by the promoter DNA sequence, the first and second RNA bases bind to the complex and RNA polymerase catalyzes formation of a covalent bond between them (thick line represents a short, released, abortive transcript). The vertical part of the arrow indicates the position on DNA where RNA synthesis commences, the horizontal part the direction of movement of RNA polymerase on the DNA. (C) Elongation. After promoter clearance, the RNA polymerase extends the growing RNA chain across the gene (the thick line is the growing RNA transcript). (D) Termination. When the elongating complex reaches the terminator, RNA synthesis stops and the complex is released from the DNA.

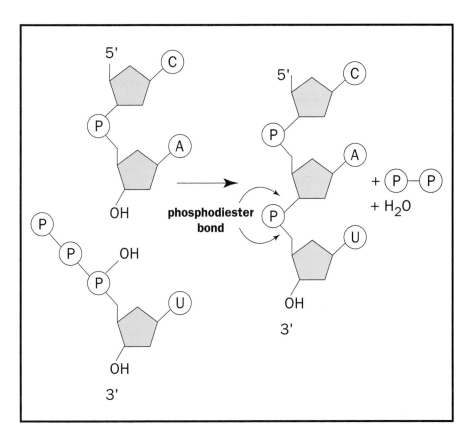

RNA nucleotides link together by dehydration, the removal of a water molecule and the bonding of the remaining atoms. The pair of P-O-C linkages connecting the two sugars is called a phosphodiester bond. The triphosphate of the incoming nucleotide provides the energy needed to carry out this reaction.

Before the nucleotides are linked together, they exist separately as ribonucleoside triphosphates (NTPs). As shown below, the NTPs contain one of the four common RNA bases, A, C, G, and U, linked to a five-carbon ribose sugar, linked, in turn, to a chain of three phosphate groups. During RNA synthesis, a covalent, "phosphodiester" bond is formed between one of the three phosphate groups on one NTP and a hydroxyl group on another. The two other phosphate groups that were part of the original NTP are released.

RNA synthesis is said to proceed in the 5′ to 3′ direction, reflecting the fact that the attachment of new nucleotides always occurs at the 3′ hydroxyl group of the growing RNA chain. RNA synthesis goes through phases that are typical of **polymerization** processes: initiation, elongation, and termination, yielding an RNA product of defined size and sequence.

polymerization linking together of similar parts to form a polymer

Initiation. The first phase of RNA synthesis is initiation (Figure 1B). Initiation starts when the first phosphodiester bond is formed. At precise locations, determined by the promotor DNA sequence, the first and second RNA bases bind to the complex, and RNA polymerase catalyzes the formation of a covalent bond between them.

When the growing RNA chain reaches a length of about ten nucleotides, the complex loses contact with the promoter and starts moving along the DNA. This is referred to as promoter "clearance" or "escape."

Only a fraction of initiation events lead to promoter clearance. In many instances, an "abortive" RNA molecule, shorter than ten nucleotides, is released from the RNA polymerase, and RNA synthesis begins all over again. Such an abortive molecule is shown in the figure as a thick line.

Once the growing RNA chain has reached the critical length of about ten nucleotides, the initiation stage is considered to have ended, and elongation begins. In eukaryotes, the transition from initiation to elongation can be triggered by enzymes called kinases, which attach phosphate groups to RNA polymerase, facilitating promoter clearance.

Elongation. Genes range in length from about 80 base pairs of DNA, as is the case for those transcribed into transfer RNA, to more than 1 million base pairs, as is the case for those encoding very long proteins. An RNA polymerase molecule that has disengaged from DNA during elongation would be unable to finish synthesizing the RNA molecule. Thus the enzyme has to traverse even the longest genes (Figure 1C), without falling off.

Along the way, there are DNA sequences that the RNA polymerase traverses considerably more slowly than at its usual rate of about 50 nucleotides per second. At regions called pause sites, it may take longer than 1 second for a single nucleotide to be added to the growing polymer.

introns untranslated portions of genes that interrupt coding regions

In eukaryotes, many genes contain blocks of DNA called **introns**, which disrupt the coding information of the gene. Introns are removed from the newly made RNA by a process called splicing. It is thought that the proteins which carry out the splicing are carried by the RNA polymerase as it is transcribing the gene, allowing the processing of the RNA to occur at the same time as the RNA molecule is synthesized.

Termination. When the RNA polymerase reaches a specific DNA sequence known as a terminator, it slows down and the transcription complex dissociates from the DNA, as shown in Figure 1D. The released RNA polymerase is then free to participate in a new initiation event.

At some terminators, primarily in bacteria, the RNA polymerase is able to respond to the release signal without being helped by any other proteins. Such sites are called intrinsic terminators. At other sites, termination is accomplished only with the aid of additional proteins. These proteins, called termination factors, are also instrumental in causing RNA to be released from the transcribing complex.

Archaea one of three domains of life, a type of cell without a nucleus

"Factor-dependent" terminators have been found in organisms from each of the three domains of life, the eukaryotes, bacteria, and **Archaea**. In eukaryotes, but usually not in bacteria, transcription of most genes proceeds past the end of the gene, as shown in Figure 1D.

The initial RNA molecules are often referred to as "primary" transcripts. In many instances, the primary transcripts must be processed to yield functional, or "mature," RNA. The processing can involve shortening them by removing their terminal or internal regions, or modifying specific nucleotides in other ways.

Regulation of Transcription

Only a few of an organism's genes are active or "expressed" at any particular time. Which genes are expressed in a particular cell depends on such factors as the nutrients available, the cell's state of differentiation, and the cell's age. There are intricate mechanisms that let the cell regulate the expression of many of its genes. Transcription, the first step in the expres-

sion of the genetic information, is an important point at which gene expression can be regulated.

There are two types of regulation: positive control, in which transcription is enhanced in response to a certain set of conditions; and negative control, in which transcription is repressed. Usually, positive control is used at promoters that are otherwise engaged in the initiation of few RNA molecules. Negative control is used at promoters where many molecules of RNA are initiated.

Activator proteins enable positive control by binding to the promoter to recruit RNA polymerase or other required initiation proteins. Such activator proteins usually bind upstream of the promoter (Figure 1). Increased recruitment then leads to an increased rate of synthesis of RNA for a particular gene. The more regulatory sites that are bound, the greater the increase in the rate of RNA synthesis. Repressor proteins can inhibit initiation of transcription by binding to the promoter and preventing RNA polymerase or a required initiation protein from binding.

In eukaryotes, DNA is "packaged" into **nucleosomes** by being wrapped around **histone** proteins. This can dramatically reduce the ability of genes to be transcribed, because the packaging may hide promoter sequences that are recognized by initiation factors.

nucleosomes chromosome structural units, consisting of DNA wrapped around histone proteins

histone protein around which DNA winds in the chromosome

Two mechanisms are used to alter the DNA packaging, to regulate transcription. First, enzymes called chromatin remodeling factors can move histone proteins around on the DNA, so that promoter sequences are more accessible or less accessible to the transcription initiation machinery. Second, enzymes can attach small chemical groups, including acetyl, phosphate, methyl or other groups, to the histone proteins. This modification of histone proteins may alter the interaction between the DNA and the histones, or between histones and other proteins, either facilitating or blocking the ability of initiation factors to bind DNA.

Transcription also is regulated by proteins that influence how quickly RNA polymerase moves along the DNA. These proteins, called regulatory elongation factors, may help the polymerase traverse pause sites, and they may facilitate elongation through packaged DNA. On the other hand, they may also facilitate the termination of transcription at specific sites. SEE ALSO ARCHAEA; GENE EXPRESSION: OVERVIEW OF CONTROL; NUCLEOTIDE; OPERON; RNA POLYMERASES; RNA PROCESSING; TRANSCRIPTION FACTORS; TRANSLATION.

David T. Auble and Pieter L. de Haseth

Bibliography

de Haseth, Pieter L., Margaret Zupancic, and M. Thomas Record Jr. "RNA Polymerase-Promoter Interaction: The Comings and Goings of RNA Polymerase." *Journal of Bacteriology* 180 (1998): 3019–3025.

Lemon, Bryan, and Robert Tjian. "Orchestrated Response: A Symphony of Transcription Factors for Gene Control." *Genes & Development* 14 (2000): 2551–2569.

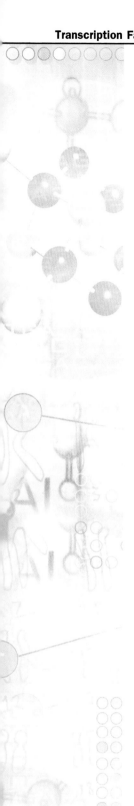

Transcription Factors

Transcription factors are protein complexes that help RNA polymerase bind to DNA. RNA polymerase is the **enzyme** that transcribes genes to make messenger RNA, which is then used to make protein. By controlling RNA polymerase's access to the gene, transcription factors control the rate at which a gene is transcribed. Without transcription factors, cells would not be able to effectively regulate the rate at which genes are expressed.

enzyme a protein that controls a reaction in a cell

Basal Transcription Factors Bind to the Gene Promoter Region

Every gene has a region known as the **promoter**. This is a DNA sequence "upstream" from the coding region, to which RNA polymerase must bind before it begins transcribing the coding region of the gene. In eukaryotes, the promoters of many (but not all) genes contain the sequence TATAA twenty-five to thirty nucleotides upstream from the transcription start site (T is the nucleotide adenine; A is thymine). Called the "TATA box," this sequence binds the TATA-binding protein (TBP), one of the most ancient and most important transcription factors.

promoter DNA sequence to which RNA polymerase binds to begin transcription

The DNA-binding region of TBP has changed very little in millions of years of evolution, indicating how central this portion of the protein is in gene transcription. Transcription factors in **archaeans** are closely akin to those in **eukaryotes**, though simpler, and they reveal a deep evolutionary relation between the two groups. (Transcription factors are also used by **eubacteria**, but the details differ significantly and will not be discussed here.)

archaeans members of one of three domains of life, have types of cells without a nucleus

eukaryotes organisms with cells possessing a nucleus

eubacteria one of three domains of life, comprising most groups previously

When TBP contacts DNA, the DNA bends. This distortion in shape allows the two sides of the double helix to come apart more easily. TATA-box DNA is especially easy to separate, because successive adenine-thymine pairs are somewhat less stable than series of other nucleotide pairs. The separation of the two strands makes the coding region of the gene more accessible to the RNA polymerase. (There are in fact three eukaryotic RNA polymerases, known as pol I, pol II, and pol III. Each uses a different set of transcription factors; we will discuss those for pol II.)

TBP plays a central role in initiating transcription, but it does not act alone. In archaeans, it works with another protein, transcription factor B. In eukaryotes, TBP is part of a larger complex, TFII-D (this rather colorless name is derived from "transcription factor D for RNA polymerase II"). TFII-D includes several other proteins besides TBP that interact with other factors and help stabilize the assembly on the DNA.

By itself, TFII-D cannot efficiently promote DNA-binding and transcription. Four other factors, TFII-B, -F, -E, and -H (binding in the order listed), allow pol II to bind to the promoter. Together, these are known as the basal transcription factors. Because each of these is composed of numerous individual polypeptides, the entire complex is thought to comprise at least twenty-five interacting polypeptides whose multiple interactions are critical for successful transcription.

The basal transcription factors assembled at the promoter are effective because they bind pol II. To finish their job, though, they must release

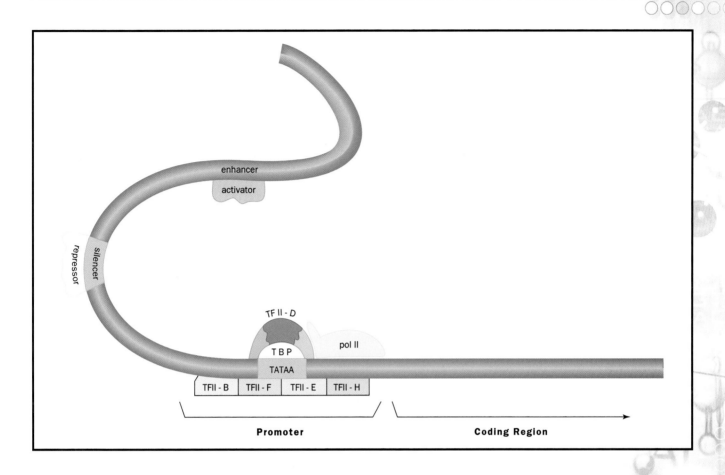

it to begin transcription. This occurs when TFII-E and -H cooperate to phosphorylate (add a phosphate group to) RNA polymerase. This changes the polymerase's shape sufficiently to allow it to escape the complex of transcription factors and begin transcription.

Gene-Specific Factors Differentially Enhance Transcription Rates

The basal transcription factors increase the rate of transcription for all genes; indeed, RNA polymerase cannot bind to the promoter without them. However, not all genes should be transcribed at an equal rate all the time. Red blood cells should make lots of hemoglobin but not the digestive enzyme pepsin, while stomach lining cells should do the opposite. Differential control of gene transcription is facilitated by gene-specific transcription factors.

Hormones are an important class of molecules that regulate gene expression. A hormone is not a transcription factor itself but binds to a receptor to form a gene-specific factor. Once bound together, the hormone-receptor complex binds to DNA. Growth factors and homeotic proteins also act as gene-specific factors or form complexes that do.

The number of known gene-specific factors is currently in the low thousands and inevitably will grow as the **genome** becomes better known. An average gene may have several dozen specific factors involved in its regulation, giving the potential for very precise control of its expression.

Schematic diagram of a gene and its regulatory regions. TFII-D, -B, -F, -E, and -H are basal transcription factors, required for binding of RNA polymerase (pol II) to the promoter. Activators and repressors enhance or reduce the rate of gene transcription.

genome the total genetic material in a cell or organism

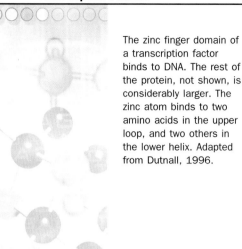

The zinc finger domain of a transcription factor binds to DNA. The rest of the protein, not shown, is considerably larger. The zinc atom binds to two amino acids in the upper loop, and two others in the lower helix. Adapted from Dutnall, 1996.

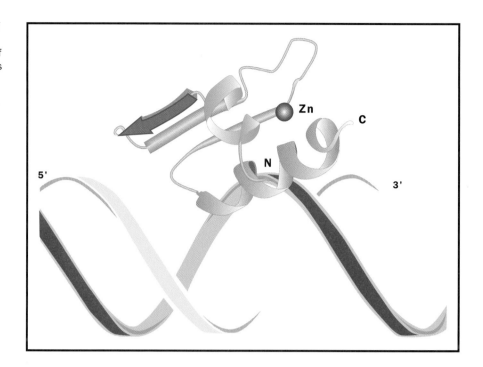

Gene-specific factors are known as activators or repressors, depending on whether they increase or decrease the rate of transcription. The DNA sequences that activators bind to are called enhancer sites; repressors bind to silencer sites. Since enhancer and silencer sites are on the same DNA sequence as the gene they control, they are called "cis" regulatory elements (from the Latin word for "side"). The factors that bind to them come from elsewhere in the genome and are called "trans" acting factors.

Many gene-specific factors bind to the promoter outside of the TATA box, especially near the transcription initiation site, the beginning of the DNA sequence that is actually read by RNA polymerase. Others bind to sequences within the coding region of the gene, or downstream from it at the termination region. Some bind to DNA sequences hundreds or thousands of nucleotides away from the promoter. Because of the looped structure of DNA, these sequences are physically close to the promoter, despite being far away along the double helix. The binding sites of transcription factors can be determined by "DNA footprinting."

Gene-specific factors work in a variety of ways. Some interact with the basal factors, altering the rate at which they bind to the promoter. Some influence RNA polymerase's rate of escape from the promoter, or its return to it for another round of transcription.

Some factors physically alter the local structure of the DNA, making it more or less accessible. In **eukaryotic** organisms, DNA is wound around protein complexes called **histones** and is further looped, coiled, and condensed to allow efficient packing in the cell nucleus. This arrangement keeps the DNA well ordered but also decreases its accessibility for transcription. By interacting directly with DNA, transcription factors can open up otherwise inaccessible regions.

eukaryotic describing an organism that has cells containing nuclei

histones proteins around which DNA winds in the chromosome

Gene-Specific Factors Share Several Common DNA-Binding Motifs

Gene-specific factors must position themselves on specific DNA sequences to exert their effects, and the relatively simple structure of DNA offers only a few ways for proteins to grab on. Therefore, despite the wealth of different individual transcription factors, each employs one of only a handful of structural "motifs" to bind to the DNA double helix. Each motif is a small portion of a much larger protein, whose other portions confer DNA-sequence specificity and control its interaction with the basal factors or other proteins.

The helix-turn-helix motif is composed of a short section of alpha-helix, linked to a loop of amino acids that changes the direction of the chain, followed by another alpha-helix. The first helix fits into the so-called major groove of the DNA double helix. The side chains of the protein's amino acids make contact with the exposed portions of the nucleotides. The shape and charges of the one complement those of the other, allowing them to bind; this provides the sequence-specificity needed for effective gene regulation. The homeotic proteins are a special class of proteins employing a modified helix-turn-helix motif. These proteins play critical roles in regulating development in organisms as diverse as fruit flies and humans.

The zinc-finger motif is constructed around an atom of zinc, which binds four amino acids to hold the amino acid chain in proper orientation. While many of the other amino acids vary among different types of zinc-finger proteins, the four key amino acids—either four cysteines or two cysteines and two histidines—are invariant in this class of transcription factors. This group of factors includes the steroid receptors. Steroids are a class of **hormones**, including testosterone and the estrogen, that exert profound effects on development. Steroids must bind to a receptor to form the transcription factor complex. **Mutations** in steroid receptors are responsible for a large variety of inherited disorders, including androgen insensitivity syndrome, thyroid hormone resistance syndrome, and some forms of prostate cancer, breast cancer, and osteoporosis.

hormones molecules released by one cell to influence another

mutations changes in DNA sequences

Regulation of Transcription Factors

All cells need to be responsive to their environments, whether that environment is the pond-water habitat of a *Paramecium* or the thousands of other cells that a single neuron communicates with every second. Transcription factors are a central feature of this responsiveness. The hormone-receptor complex mentioned above provides a model for understanding how a cell can coordinate its gene expression with external events. The hormone acts as a signal that a change has occurred in the outside world that requires action by the cell, whether it be to grow or divide, or to release its own hormone.

A key feature governing a cell's repertoire of responses is the set of receptors it makes. Cells that should not respond to testosterone need only ensure that they do not make the testosterone receptor—a decision itself governed by the presence or absence of other transcription factors. Hormones are not the only type of signal possible. Cells have complex networks of signaling pathways that help to regulate their actions.

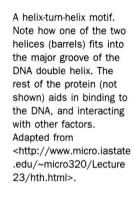

A helix-turn-helix motif. Note how one of the two helices (barrels) fits into the major groove of the DNA double helix. The rest of the protein (not shown) aids in binding to the DNA, and interacting with other factors. Adapted from <http://www.micro.iastate.edu/~micro320/Lecture 23/hth.html>.

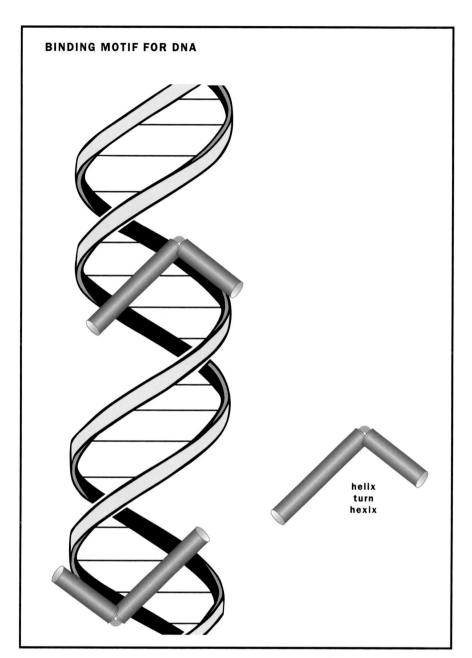

BINDING MOTIF FOR DNA

helix
turn
hexix

The exquisite coordination of cellular processes needed to maintain life might be likened to a symphony, in which the many different instruments must play their parts in time with all the others. The timing of gene expression is one of the great puzzles of understanding life: How does each gene get turned on and off at the right time? Although transcription factors are clearly an important part of the answer, they themselves are proteins—the products of genes that must be regulated by yet other transcription factors.

The way out of this paradox is to remember that each organism does not arise from nothing, nor does it spring forth fully formed, with all of its parts fully functioning. Rather, it develops from a preexisting cell, with a specific set of transcription factors in place to turn on a specific set of developmental genes, many of which are themselves transcription factors that

turn on other genes. Although this is only the barest outline of an explanation that is still being worked out, it is clear that the pulsing interplay of transcription factors is a central feature of life's coordinated complexity. SEE ALSO DEVELOPMENT, GENETIC CONTROL OF; DNA FOOTPRINTING; GENE EXPRESSION: OVERVIEW OF CONTROL; HORMONAL REGULATION; PROTEINS; RNA POLYMERASES; SIGNAL TRANSDUCTION; TRANSCRIPTION.

Richard Robinson

Bibliography

Alberts, Bruce, et al. *Molecular Biology of the Cell,* 4th ed. New York: Garland Science, 2002.

Semenza, Gregg L. *Transcription Factors and Human Disease.* Oxford, U.K.: Oxford University Press, 1998.

Weinzierl, Robert O. J. *Mechanisms of Gene Expression: Structure, Function and Evolution of the Basal Transcriptional Machinery.* London: World Scientific, 1999.

Internet Resources

Dutnall, Robert N., David N. Neuhaus, and Daniela Rhodes. "The Solution Structure of the First Zinc Finger Domain of SWI5: A Novel Extension to a Common Fold." *Structure* 4 (1996): 599–611. <http://baldrick.ucsd.edu/~dutnall/StructureDisplay.html>.

TRANSFAC: The Transcription Factor Database. GBF-Braunschweig. <http://transfac.gbf.de/TRANSFAC/>.

Transduction

Transduction is one of three basic mechanisms for genetic exchange in bacteria. Like transformation and conjugation, transduction allows the movement of genetic information from a donor cell to a recipient. Unlike the other mechanisms, however, transduction requires the participation of a type of virus called a bacteriophage in order to accomplish this movement. While transduction has been studied in the laboratory since the 1950s, more recently scientists have shown that the process also occurs in nature and probably plays an important role in the evolution of bacteria.

While transduction is common to many bacteria (but not all), the processes can be divided into two basic mechanisms. Generalized transduction tends to transfer all bacterial genes with similar frequencies, or number of cells genetically altered as a function of the total number of potential recipient cells. Specialized transduction tends to transfer only specific genes. It is the life cycle of the particular virus involved in transduction that determines which mechanism will occur, because the transduction of bacterial chromosomal genes is, in fact, the result of an error in the mechanism for viral replication. Thus, to understand the processes of transduction, one must understand the basic mechanisms of viral replication.

Generalized Transduction

Generalized transduction is usually mediated by certain lytic viruses. A lytic virus is one that normally infects a host cell and redirects the resources of the host away from its own cellular replication toward viral replication and the eventual lysis, or breaking open, of the bacterial cell. Typical lytic viruses

LYTIC CYCLE

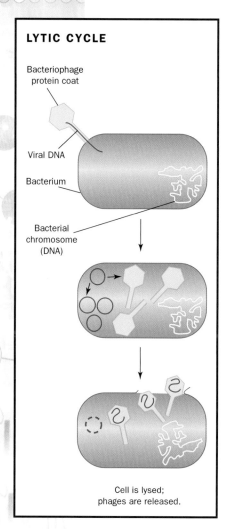

Bacteriophage
protein coat

Viral DNA

Bacterium

Bacterial
chromosome
(DNA)

Cell is lysed;
phages are released.

In transduction, a virus inserts its genetic material into a bacterial cell, as shown, but then integrates it into the bacterial chromosome. When the viral DNA leaves the bacterial chromosome, it may take one or more bacterial genes with it, packaging them into viral protein coats in which they are released from the cell.

transcription messenger RNA formation from a DNA sequence

translation synthesis of protein using mRNA code

nuclease enzyme that cuts DNA or RNA

use the **transcription** and **translation** mechanisms of the host cell to express viral genes.

Viral genes are expressed in a specific order. First, the genes needed to hijack host cell metabolism are expressed. Next, the viral genetic information is copied many times. The third step involves transcription and translation of viral genes for protein components. The fourth step requires the insertion of viral DNA into the capsid, the protein shell of the virus. Finally, the virus expresses genes needed to break open the host cell, releasing hundreds of new viral particles that move on to infect new host cells.

Among the best studied of the generalized transducing viruses is P22, a virus of the bacterium *Salmonella choleraesuis* (subspp. *typhimurium*). In the course of its normal life cycle, the virus binds to a specific receptor on the surface of the bacterial cell, injects the viral DNA into the host cell, and begins to express viral genes. The ends of the viral DNA are redundant; that is, they have small sections in which the genetic information is duplicated. The DNA repair enzymes of the host cell joins these ends so that the viral DNA forms a circle of double-stranded DNA. A DNA **nuclease** enzyme then makes a nick in one of the strands of DNA. The unbroken strand then serves as a template, allowing the broken strand to extend itself. This forms a long stretch of DNA that repeats the genome of the virus many times.

The long repeating sequence of DNA is known as a concatamer. The genes for viral capsid proteins (the proteins that make up the coat of the virus) are then expressed, and empty viral capsids are produced. Next, a viral DNA-cutting enzyme cuts the concatamer of DNA at a specific sequence. Once the concatamer is cut, the end of the DNA is pushed into an empty viral capsid until the viral head is full. A new empty capsid then positions itself on the end of the concatamer and the process is repeated until all the viral DNA is packaged.

Generalized transduction occurs when the enzyme that normally cuts the concatamer cuts the host chromosome instead. The viral capsids cannot distinguish viral DNA from host DNA; thus, once the initial cut is made, the empty capsids are filled with host chromosomal DNA instead of viral DNA. Since each empty capsid picks up where the last ended, it is possible for all the genes on the host chromosome to be packaged at similar rates: thus, this process is generalized. When the host cell breaks open, the viral capsids containing host chromosomal DNA, now called transducing particles, are able to bind to a new host cell. The DNA they carry is injected into the new host, but since there are no viral genes included on the injected DNA, the newly infected host cell is not killed. If the original host cell (the donor) was genetically different from the new host cell (the recipient), the DNA recombination enzymes of the recipient cell will insert the new genes in place of the old, thus altering its own genetic makeup. This genetic change is integrated into the recipient genome and is passed on to future progeny cells.

Specialized Transduction

Specialized transduction results in the movement of only specific genes. The viruses that carry out specialized transduction are called lysogenic viruses. They have a mechanism for replication that is different from that of generalized transduction, for they integrate their DNA directly into the chromosome of its host's genome. Each time the host chromosome is duplicated, so is the integrated viral DNA. In many cases these viruses express genes that keep the viral DNA dormant; that is, the virus does not immediately replicate.

The best characterized model for a specialized transducing virus is phage Lambda of *Escherichia coli*. The Lambda virus begins its life cycle in much the same way as P22. A tail fiber on the end of the virus specifically binds to a receptor, called the maltose binding protein, on the surface of the *E. coli* cell. The viral DNA is then injected into the host cell. On the chromosome of the virus is a section of DNA that is almost identical to a DNA sequence found on the bacterial chromosome. The recombination enzymes of the *E. coli* host break and rejoin the viral DNA and host DNA together at this site, thus integrating the virus genome into the chromosome of the host bacterium.

A viral gene, the repressor gene, is then expressed and keeps the virus from activating its own replication. This integrated virus, called a prophage, can be maintained stably as a part of the host chromosome as long as the host cell remains healthy. If the host cell becomes damaged, enzymes are activated that destroy the Lambda repressor protein. Without this protein, the viral DNA will break out of the host chromosome and begin to replicate itself, much as a lytic virus does. Ultimately, the virus particles are packaged, released, and move on to infect a new host cell.

Specialized transduction occurs when the enzyme that cuts the viral DNA out of the host chromosome makes a mistake and cuts in the wrong place, removing some, but not all, of the viral genes. Since Lambda capsids fill by a "headful" mechanism, small bits of the host chromosome are packaged along with part of the viral genes. These viral particles are called defective particles because some of the viral genes are missing in the package and thus, when the virus infects a new host cell, not all the genes needed for viral replication are present. Lacking the ability to replicate, the virus cannot kill its new host. Because of this mechanism of viral packaging, specialized transducing viruses can pick up genes only on either side of the site where the virus integrates into the bacterial chromosome. Thus, while generalized transducing viruses can move any genes, specialized transducing viruses move only specific genes.

Uses in Research

Generalized transducing viruses are the most useful in mapping bacterial chromosomal genes. Since the amount of DNA that is packaged by the virus is determined by the size of the head of the virus, each viral particle holds the same amount of DNA. The initial cutting of the host chromosome is a random event, giving all genes approximately the same probability of being packaged and transferred. Each piece of DNA that is packaged will be the same length, meaning that the closer together two genes are, the higher the probability that the two genes will be present on the same fragment of packaged DNA. In other words, the closer together the genetic markers are, the higher the frequency of cotransduction. Therefore the distance between closely linked chromosomal genes can be calculated by measuring the frequency that two genes or genetic markers are cotransduced.

When the distance between two genes is greater than the size of the viral genome, it is physically impossible for the two genes to be packaged in the same viral capsid. Thus, these genes are said to be "unlinked" with regard to viral mapping. Since most transducing viral capsids can hold only from about fifteen to fifty genes, transductional mapping of bacterial chromosomal genes is most effective for genes that are relatively close to one another.

Historically, viruses, including transducing viruses, have played an important role in defining the basic principles of molecular biology. Perhaps the most important contribution to the study of transduction was that made by Alfred Hershey and Martha Chase. During the 1940s and 1950s there was still a great deal of controversy over whether DNA or protein was the genetically inheritable material. Hershey and Chase recognized that the simplicity of the virus, consisting of DNA wrapped in a protein coat, was the ideal model to directly address the question of the basis for inheritance.

media (bacteria) nutrient source

They began their experiments by growing viruses on host bacteria in **media** containing radioactive forms of sulfur and phosphorus. The radioactive sulfur labeled the protein components of the virus, while the radioactive phosphorus labeled the DNA portions. This allowed them to independently track the protein and DNA. After separating the radioactively labeled viruses from their host cells, they used the viruses to infect host bacteria that were not radioactively labeled. After infection, they separated the bacterial cells from the growth media. Radioactive phosphorus (viral DNA) was found inside the host cells, while the radioactive sulfur (viral proteins) was found outside the cell. This indicated that only the DNA of the virus enters the host cell, while the protein was left on the outside.

Further details of their experiment make the case for DNA even more strongly. When viruses are grown on bacteria in a thin layer on the surface of agar plates, each virus will create a clear spot called a plaque, indicating infection. Hershey and Chase had two different mutants of their virus that resulted in plaques that looked different from the normal viral infection.

When the infected bacteria were replated, normal plaques were seen, indicating that the two different new mutants had both infected the same host cell and that recombination between the virus DNA occurred within, making a virus that had repaired both mutations. Consequently, since only DNA had entered the host cells and genetic change had occurred in the viruses, DNA had to be the inheritable material. Proteins could not be the source of inheritance because the viral proteins never entered the host cells. SEE ALSO CONJUGATION; DNA REPAIR; *ESCHERICHIA COLI* (*E. COLI* BACTERIUM); EUBACTERIA; MAPPING; NATURE OF THE GENE, HISTORY; RECOMBINANT DNA; TRANSFORMATION; VIRUS.

Gregory Stewart

Bibliography

Curtis, Helen, and N. Susan Barnes. *Invitation to Biology*, 5th ed. New York: Worth Publishers, 1994.

Ingraham, John, and Catherine Ingraham. *Introduction to Microbiology*, 2nd ed. Pacific Grove, CA: Brooks/Cole Publishing, 1999.

Madigan, Michael T., John Martinko, and Jack Parker. *Brock Biology of Microorganisms*, 10th ed. Upper Saddle River, NJ: Prentice Hall, 2000.

Streips, Uldis N., and Ronald E. Yasbin. *Modern Microbial Genetics*, 2nd ed. Hoboken, NJ: John Wiley & Sons, 2002.

Transformation

Transformation is one of three basic mechanisms for genetic exchange in bacteria. Transformation may be either a natural process—that is, one that has evolved in certain bacteria—or it may be an artificial process whereby the recipient cells are forced to take up DNA by a physical, chemical, or enzymatic treatment. In both cases, **exogenous** DNA (DNA that is outside the host cell), is taken into a recipient cell where it is incorporated into the recipient **genome**, changing the genetic makeup of the bacterium.

exogenous from outside

genome the total genetic material in a cell or organism

Natural Transformation

Natural transformation is a physiological process that is genetically encoded in a wide range of bacteria. Most bacteria must shift their physiology in order to transform DNA; that is, they must become "competent" for taking up exogenous DNA. There appear to be two basic mechanisms by which bacteria can become competent for transformation. In some bacteria, including *Streptococcus pneumoniae* and *Bacillus subtilis*, competence is externally regulated. These bacteria produce and secrete a small protein called competence factor that accumulates in the growth medium.

When the bacterial culture reaches a sufficient density, the concentration of competence factor reaches a level high enough to bind receptors on the outside of the cell. This event causes an internal signal to turn on the expression of the genes needed for transformation. Thus, competence development is controlled by cell density. There are a number of other bacterial functions that are similarly regulated, and these processes are collectively called quorum sensing mechanisms. In other bacteria, including *Haemophilus influenzae* and *Pseudomonas stutzeri*, competence development is internally regulated. When there is a shift in the growth dynamics of the bacterium, an internal signal triggers competence development.

Once competence is induced, three additional steps are required for natural transformation. After induction of competence, double-stranded DNA is bound to specific receptors on the surface of the competent cells. These receptors are lacking in noncompetent cells. The double-stranded DNA is nicked and one strand is degraded while the other strand enters the cell. This process is called DNA uptake. Finally, the recombination enzymes of the recipient cell will bind the single-strand DNA that has entered it, align it with its **homologous** DNA on the recipient chromosome, and recombine the new DNA into the chromosome, incorporating any genetic differences that exist on the entering DNA.

homologous containing the same set of genes

Artificial Transformation

While a wide variety of bacteria can transform naturally, many species cannot take up DNA from an outside source. In some cases DNA can be forced into these cells by chemical, physical, or enzymatic treatment. This is especially important in genetic engineering, as artificial transformation is essential for the introduction of genetically altered sequences into recipient cells. One of the two most common methods is a chemical process where cells are heat-shocked, then treated with the DNA and a high concentration of calcium ions. The calcium ions precipitate the DNA on the surface of the cell, where the DNA is forced into the recipient.

In transformation, bacteria pick up DNA from their environment. This illustration shows Frederick Griffith's classic experiment which first demonstrated transformation. The live nonvirulent bacteria absorb DNA from the dead nonvirulent bacteria, and become virulent themselves. Adapted from Curtis and Barnes, 1994.

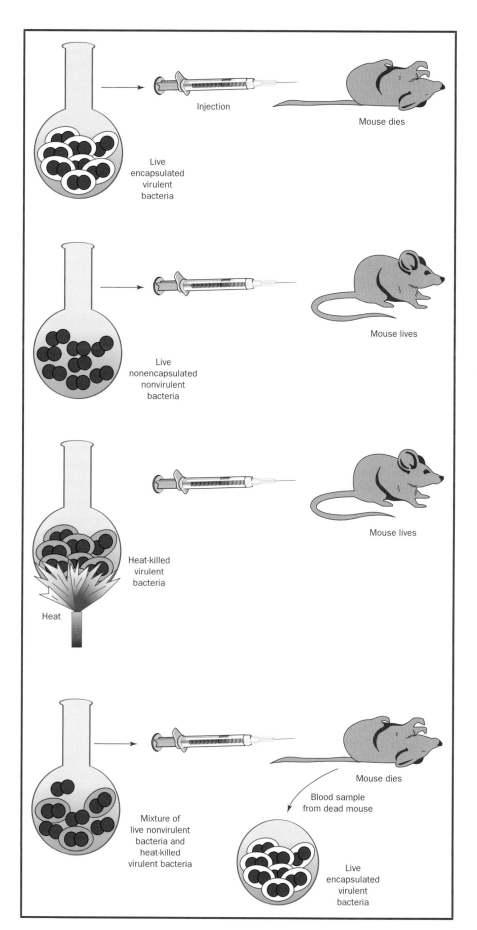

Injection

Live encapsulated virulent bacteria

Mouse dies

Live nonencapsulated nonvirulent bacteria

Mouse lives

Heat-killed virulent bacteria

Heat

Mouse lives

Mixture of live nonvirulent bacteria and heat-killed virulent bacteria

Mouse dies

Blood sample from dead mouse

Live encapsulated virulent bacteria

More recently a new method, called electroporation, has been used to introduce DNA by artificial transformation. In this process a suspension of recipient bacteria and transforming DNA is placed in a container with metal sides. A high-voltage electrical current is passed through the sample, temporarily creating small pores, or channels, in the membranes of the bacteria. The DNA enters the cells and the pores close. Thus, exogenous (outside) DNA is introduced into the recipient.

Because exogenous DNA is not enclosed within cell walls, it is susceptible to enzymes that degrade DNA, called DNases. A hallmark of transformation is that it is sensitive to DNase, while the other two processes of genetic exchange, **transduction** and **conjugation**, are DNase resistant. Transduction is DNase resistant because the DNA is protected inside a viral protein coat. Conjugation is DNase resistant because fusion occurs between donor and recipient cells, meaning the DNA is never exposed to the outside environment or to enzymes.

transduction conversion of a signal of one type into another type

conjugation a type of DNA exchange between bacteria

Discovery of Transformation

The first report of transformation was an example of natural transformation. Dr. Frederick Griffith was a public health microbiologist studying bacterial pneumonia during the 1920s. He discovered that when he first isolated bacteria from the lungs of animals with pneumonia, the bacterial colonies that grew on the agar plates were of reasonable size and had a glistening, **mucoid** appearance. When he transferred these colonies repeatedly from one agar plate to another, however, mutant colonies would appear that were much smaller and were chalky in appearance. He designated the original strains as "smooth" strains, and the mutants as "rough" strains. When Griffith injected mice with smooth strains they contracted pneumonia, and smooth strains of the bacterium could be reisolated from the infected mice. However, when he infected the mice with rough strains they did not develop the disease. The smooth strains were capable of causing disease, or were "virulent," while the rough strains did not cause disease, or were "aviruluent."

mucoid having the properties of mucous

Griffith questioned whether the ability to cause disease was a direct result of whatever product was making the bacterial colonies smooth, or whether rough strains of the bacterium were less capable of establishing disease for some other reason. To investigate this idea, he prepared cultures of both bacterial types. He pasteurized (killed) each of these cultures by heating them for an hour and then injected the heat-treated extracts into mice. His hypothesis was that if the bacteria had to be living to cause disease, heat-treating that killed the bacteria would prevent disease. If, on the other hand, the smooth material was itself a toxin, heating would not destroy it, meaning heated extracts of smooth strains would continue to cause disease. When Griffith injected heated extracts of both smooth and rough strains into mice, neither caused disease. This suggested to him that only living smooth cells could cause disease.

In his next experiment he coinjected unheated, live rough bacteria with heat-treated, dead smooth bacteria into mice. All of the mice developed disease, and when bacteria were isolated from the lungs of the diseased mice, all the isolates were smooth. This led Griffith to propose that there was some "transforming principle" in the heated smooth extract that con-

verted the rough strains back to smooth ones capable of causing diseases. Griffith was not able to determine the nature of this transforming principle, but his experiments suggested that some "inheritable" material present in the heated extract could genetically convert strains from one colony type to another.

Approximately ten years later, another research team, that of Oswald Avery, Colin Munro MacLeod, and Maclyn McCarty, followed up on Griffith's experiments by enzymatically and biochemically characterizing the heated transforming extracts that Griffith had produced. Their studies indicated that the transforming principle was deoxyribonucleic acid (DNA), providing the first definitive evidence that DNA was the inheritable material. SEE ALSO CONJUGATION; NATURE OF THE GENE, HISTORY; RECOMBINANT DNA; TRANSDUCTION.

Gregory Stewart

Bibliography

Curtis, Helen, and N. Susan Barnes. *Invitation to Biology*, 5th ed. New York: Worth Publishers, 1994.

Ingraham, John, and Catherine Ingraham. *Introduction to Microbiology*, 2nd ed. Pacific Grove, CA: Brooks/Cole Publishing, 1999.

Madigan, Michael T., John Martinko, and Jack Parker. *Brock Biology of Microorganisms*, 10th ed. Upper Saddle River, NJ: Prentice Hall, 2000.

Streips, Uldis N., and Ronald E. Yasbin. *Modern Microbial Genetics*, 2nd ed. Hoboken, NJ: John Wiley & Sons, 2002.

Transgenic Animals

genome the total genetic material in a cell or organism

The term "transgenics" refers to the science of inserting a foreign gene into an organism's **genome**. Scientists do this, creating a "transgenic" organism, to study the function of the introduced gene and to identify genetic elements that determine which tissue and at what stage of an organism's development a gene is normally turned on. Transgenic animals have also been created to produce large quantities of useful proteins and to model human disease.

In the early 1980s Frank Ruddle and his colleagues created the first transgenic animal, a transgenic mouse. Researchers making transgenic mice use a very fine glass needle to inject pieces of DNA into a fertilized mouse egg. They inject the DNA into one of the egg's two **pronuclei**, before the pronuclei fuse to become the nucleus of the developing embryo's first cell. After the DNA is injected, multiple copies, usually joined end-to-end, insert randomly into the host organism's nuclear DNA.

pronuclei egg and sperm nuclei before they fuse during fertilization

Multiple injected embryos are then transferred to a surrogate mother mouse to develop to term. Only a small percentage of the embryos survive the injection, and even of those that survive, not all have successfully incorporated the foreign DNA into their genome. Once the mice are born, researchers must identify which mice have the foreign gene in their genome. The animals that contain the added foreign DNA, or transgene are referred to as transgenics.

Targeted Gene Replacement and "Knockouts"

A gene that is injected into a fertilized egg is generally integrated randomly into the host genome. This means that scientists originally had no control over where in the host genome the foreign gene would land, nor could they control the number of copies of the gene that would be integrated. Where and how many copies of a gene are inserted can profoundly affect its function, so scientists looked for ways to make more precise insertions.

In the late 1980s Mario Capecchi and colleagues pioneered a method to target the inserted gene to a desired position in the genome. These researchers took advantage of an observation that, on rare occasions, an injected, mutated copy of a gene lines up precisely with the original form of the gene in the mouse genome. By a process called homologous recombination, the aligned DNA segments are cut and rejoined to each other. The result is a precise stitching of the introduced DNA into the targeted gene in the mouse genome. This means that scientists found they could make minor modifications to a gene before injecting it and, by homologous recombination, or "gene targeting," replace the natural gene with this transgenic version. For commercial applications, the transgene is often a gene coding for a functional human protein, which is then mass-produced in the host organism and isolated. For research purposes, it is often more useful to insert a mutated, nonfunctional version of the gene, to see what happens when the normal, functional version is missing. Creating such "knockout" organisms is a key tool used for studying genes that control development.

Chimera generated by injecting RW4 ES cells. Black mouse is a littermate control.

Selection of Gene Targeted Cells

Homologous recombination is a very rare event, and scientists using it to modify or "knock out" mouse genes must identify the cells in which it has occurred. In addition to injecting the gene they are trying to incorporate, scientists also inject "selectable" genes whose products permit cells to live or cause them to die in the presence of a particular drug. The two most common selectable genes used in gene targeting are the neomycin resistance (*neo*^r) gene, which allows cells to survive in the presence of the antibiotic neomycin, G418, and the thymidine kinase (*TK*) gene from the herpes virus. Cells with this gene die in the presence of the antiviral agent gancyclovir. The *neo*^r and *TK* genes are generally used together for maximum selection.

The first step in gene targeting is to clone the gene that is to be replaced from the mouse genome. The cloned gene is placed into a targeting vector along with a selectable gene such as the *neo*^r gene. (The targeting vector is a larger piece of carrier DNA.) When the targeting vector lines up with the native mouse gene and homologous recombination stitches the genes in the targeting vector into the genome, the *neo*^r gene will be included. By adding G418 to the cell growth media, only those cells that have incorporated into their genome the transgenes, including the *neo*^r gene, will survive. This is referred to as positive selection, selecting for cells that contain the desired integration product.

The *TK* gene is placed in the same targeting vector, but it is placed outside of the cloned mouse gene pieces. If homologous recombination occurs such that the added DNA lines up precisely with the native gene, the *TK* gene will be excluded. However, if the targeting vector integrates randomly

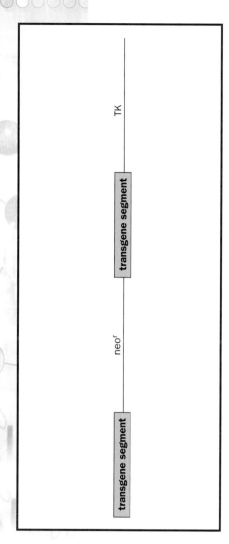

Targeting vector.

blastocyst early stage of embryonic development

into the genome the entire vector is inserted and the *TK* gene will be included. The addition of gancyclovir kills all cells that have the *TK* gene, thereby killing those where the insertion was random. The combination of positive *neo*ʳ selection and negative *TK* selection results in the survival of only those cells containing targeted gene replacements.

From Transgene to Transgenic Organism

To get the targeted gene into the mouse genome, Capecchi used a very specialized embryonic cell previously isolated by Matthew Kaufman and Martin Evans. These embryonic stem (ES) cells were isolated from an early stage of embryonic development (the **blastocyst**). When grown under controlled conditions in culture dishes, the ES cells have the remarkable capacity to become cells belonging to any tissue type. They can become muscle, cartilage, blood vessel or nerve cells, for example. Even more astonishing, when ES cells are injected back into a blastocyst, they mix with the cells of the recipient embryo and contribute some cells to every tissue in the body. Thus, if researchers place a transgene into the ES cell's genome and inject those ES cells into a blastocyst, the transgene could end up in every tissue type of the mouse.

Typically, the targeting vector is placed into ES cells derived from a mouse that has a brown coat. ES cells with the targeted gene replacement are identified through positive and negative selection. They are then injected into blastocysts from mating mice with black coats. These blastocysts are then placed into a surrogate mother and allowed to develop to term. The offspring that have incorporated transgene-containing ES cells into some of their tissues are identified by having patches of brown coat color.

To allow the targeted gene replacement to be passed to subsequent generations, the ES cells must also contribute to the developing embryo's eggs or sperm. To determine if the targeted gene has been incorporated into a mouse's eggs or sperm, the mice with brown patches are mated to black-coated individuals. The brown coat color is a dominant trait, so any offspring with brown coats can be assumed to have arisen from germ cells that derived from the manipulated ES cells and thus contain the targeted gene replacement. These mice are mated to brown-coated siblings to produce homozygous transgenics, which are identified by determining if the offspring contain two copies of the transgene replacing both normal copies of the gene in the genome. These homozygous mutants are studied to look for phenotypic changes due to the transgene.

Applications

Gene targeting has been used to identify the function of hundreds of mouse genes. One dramatic example was the deletion of the *Lim-1* gene by Richard Behringer and colleagues. The mice carrying this deletion died during embryonic development because of a complete lack of brain and head structure development. This demonstrated that the *Lim-1* gene was critical for head development.

The gene responsible for sex determination in mice was also identified thanks to the use of transgenic animals. When a gene called *Sry* (sex-determining region of the Y chromosome) was microinjected into mouse embryos, the resulting transgenic mice were all male. Indeed, even in the mice that had two X chromosomes and were thus genetically female, the presence of the *Sry* gene was sufficient to cause them to develop testes and led to complete sex reversal. This clearly demonstrated that the *Sry* gene alone was responsible for sex determination.

Gene targeting is also being exploited by scientists to create models of human disease. For instance, **mutations** have been made in the mouse version of the cystic fibrosis transmembrane conductance regulator gene. Although mice with the mutated gene do not develop the devastating symptoms of cystic fibrosis in their lungs, they do develop the intestinal and pancreatic duct defects associated with the disease and thereby provide a model to study at least part of the disease. Transgenic mice overexpressing the amyloid precursor protein form deposits in the brain that resemble the amyloid plaques found in Alzheimer's patients. Mouse models such as these can potentially be used to test drug therapies and to learn more about the progression of the disease.

mutations changes in DNA sequences

One important application of transgenic technology is the generation of transgenic livestock as "bioreactors." Key human genes have been introduced into sheep, cows, goats, and pigs so that the human protein is secreted into the milk of the transgenic animal. In theory, large quantities of the human protein can be produced in the animal's milk and subsequently purified for use in medical therapies. An early example of this technology by John Clark and colleagues was the production of transgenic sheep expressing the human blood-clotting factor IX needed by many patients with hemophilia. These researchers placed the human factor IX gene under the control of a piece of sheep DNA that normally turns on the beta-lactoglobulin gene in the mammary tissue. Though the sheep secreted factor IX into their milk, the levels of the protein were very small. With advances in the efficiency of creating and expressing genes in transgenic farm animals, therapeutic proteins can now be isolated. SEE ALSO AGRICULTURAL BIOTECHNOLOGY; ALZHEIMER'S DISEASE; BIOTECHNOLOGY; CLONING GENES; CLONING ORGANISMS; CYSTIC FIBROSIS; EMBRYONIC STEM CELLS; GENE TARGETING; HEMOPHILIA; RECOMBINANT DNA; RODENT MODELS; SEX DETERMINATION; TRANSGENIC MICROORGANISMS; TRANSGENIC ORGANISMS: ETHICAL ISSUES; TRANSGENIC PLANTS; Y CHROMOSOME.

Suzanne Bradshaw

Bibliography

Capecchi, Mario R. "Targeted Gene Replacement." *Scientific American* (March 1994): 52–59.

Velander, William H., Henryk Lubon, and William N. Dorhan. "Transgenic Livestock as Drug Factories." *Scientific American* (January 1996): 70–74.

Watson, James D., Michael Gilman, Jan Witowski, and Mark Zoller. *Recombinant DNA*. New York: Scientific American Books, 1992.

Transgenic Microorganisms

A transgenic microorganism is a microbe, usually a bacterium, into which genetic information has been introduced from the outside and which possesses the ability to pass that information on to subsequent generations in a

stable manner. This is not an entirely novel idea in microorganisms, since bacteria have been practicing and perfecting this art over billions of years of evolution. We, on the other hand, have only recently learned to duplicate this phenomenon and turn it to our own purposes. Genetic engineering is the field that has developed as a consequence of research into this process. Its commercial application forms the basis of the biotechnology industry today.

Moving Genes between Species

The process by which scientists introduce new genetic material into a microorganism is called molecular or gene cloning. It involves the isolation of DNA from a source other than the microorganism itself. Source organisms span the world of living things, from microbes to plants to animals, including humans. Scientists obtain source DNA in several different ways: by disrupting cells of the target microbe (or plant or animal) and fragmenting it into small pieces, by synthesizing it from an RNA template using an enzyme called reverse transcriptase, or by knowing the specific gene sequence and synthesizing it directly in the laboratory.

Once obtained, the pieces of DNA are inserted into a small genetic component that has the ability to make copies of itself (replicate) independently from the microbial genome. This self-replicating unit is called a cloning vector. Although these genetic elements exist naturally in the form of **plasmids** and bacterial viruses, many of the ones used today have been altered to improve their properties for transferring genes. **Restriction enzymes**, which nick the donor DNA and the cloning vector at specific sites, and DNA ligase, which attaches the donor DNA to the cloning vector, allow the source genes of interest to be inserted into the cloning vector without disrupting its ability to replicate.

plasmids small rings of DNA found in many bacteria

restriction enzymes enzymes that cut DNA at a particular sequence

The next step in the process is the introduction of the cloning vector with its segment of new DNA into a living cell. Bacteria have the ability to transport DNA into their cells in a process called transformation, and this ability is commonly exploited to achieve this goal. Getting the DNA into the cell, however, is only the beginning. No transformation is 100 percent efficient, and so the bacteria that receive the gene(s) of interest must be separated from those that did not. One of the best studied and most commonly used cloning vectors, pBR322, is especially useful for this purpose, as it contains several genes for antibiotic resistance. Hence, any cell transformed with DNA containing pBR322 will be antibiotic resistant, and thus can be isolated from similar cells that have not be so transformed by merely growing them in the presence of the appropriate drugs. All that remains is to identify bacteria that are producing the product of the desired gene(s), and cloning is a success.

nucleotides the building blocks of RNA or DNA

The introduction of human genes into bacteria has several complicating wrinkles that make cloning them even more challenging. For example, a bacterial gene codes for a protein from start to finish in one long string of **nucleotides**, whereas human cells have stretches of noncoding nucleotides called introns within their genes. Bacteria do not have the same ability as human cells to remove these introns when producing proteins from the gene, and if the introns are not removed, the intended protein cannot be produced. This, along with other complications, has been overcome using many of the tools of genetic engineering.

Commercial Application

Transgenic microbes have many commercial and practical applications, including the production of mammalian products. A company called Genentech was among the earliest and most successful commercial enterprises to use genetically engineered bacteria to produce human proteins. Their first product was human insulin produced by genetically engineered *Escherichia coli*. A variety of other human **hormones**, blood proteins, and immune modulators are now produced in a similar fashion, in addition to vaccines for such infectious agents as hepatitis B virus and measles.

hormones molecules released by one cell to influence another

Another promising application of genetically engineered microbes is in environmental cleanup, or biomediation. Scientists have discovered many naturally occurring genes that code for enzymes that degrade toxic wastes and wastewater pollutants in bacteria. Examples include genes for degrading chlorinated pesticides, chlorobenzenes, naphthalene, toluene, anilines, and various hydrocarbons. Researchers are using molecular cloning to introduce these genes from several different microbes into a single microbe, creating "super microbes" with the ability to degrade multiple contaminants.

Ananda Chakrabarty created one of the first microbes of this nature in the early 1970s. He introduced genes from several different bacteria into a strain of *Burkholderia cepacia*, giving it the ability to degrade toxic compounds found in petroleum. This microbe offered a potential alternative to skimming and absorbing spilled oil. Chakrabarty's genetically modified bacterium has never been used, however, due to public concerns about the release of genetically engineered microbes into the environment. The microbe did, on the other hand, play an important role in establishing the biotechnology industry. The U.S. Patent Office granted Chakrabarty the first patent ever for the construction and use of a genetically engineered bacterium. This established a precedent allowing biotechnology companies to protect their "inventions" in the same way chemical and pharmaceutical companies have done in the past. SEE ALSO BIOREMEDIATION; *ESCHERICHIA COLI* (*E. COLI* BACTERIUM); GENE; PLASMID; REVERSE TRANSCRIPTASE; TRANSFORMATION; TRANSGENIC ANIMALS; TRANSGENIC ORGANISMS: ETHICAL ISSUES; TRANSGENIC PLANTS.

Cynthia A. Needham

Bibliography

Glick, Bernard R., and Jack J. Pasternak. *Molecular Biotechnology: Principles and Applications of Recombinant DNA*, 2nd ed. Washington, DC: ASM Press, 1998.

Madigan, Michael T., John M. Martinko, and Jack Parker. *Brock Biology of Microorganisms*, 9th ed. Upper Saddle River, NJ: Prentice Hall, 2000.

Needham, Cynthia A., Mahlon Hoagland, Kenneth McPherson, and Bert Dodson. *Intimate Strangers: Unseen Life on Earth*. Washington, DC: ASM Press, 2000.

Snyder, Larry, and Wendy Champness. *Molecular Genetics of Bacteria*. Washington, DC: ASM Press, 1997.

Transgenic Organisms: Ethical Issues

A transgenic organism is a type of genetically modified organism (GMO) that has genetic material from another species that provides a useful trait.

ecosystem an ecological community and its environment

For instance, a plant may be given genetic material that increases its resistance to frost. Another example would be an animal that has been modified with genes that give it the ability to secrete a human protein.

Bioethics addresses the impact of technology on individuals and societies. Bioethical issues include an individual's right to privacy, equality of access to care, and doctor-patient confidentiality. In the case of transgenic organisms, a major bioethical issue is freedom of choice. Yet broader issues also arise, such as the ethics of interfering with nature, and effects of transgenic organisms on the environment.

The changes that are possible with transgenesis transcend what traditional gardening or agriculture can accomplish, although these too interfere with nature. A transgenic tobacco plant emits the glow of a firefly, and a transgenic rabbit given DNA from a human, a sheep, and a salmon secretes a protein hormone that is used to treat bone disorders. If mixing DNA in ways that would not occur in nature is deemed wrong, then transgenesis is unethical. Said a representative of a group opposed to GMOs in New Zealand at a government hearing, "To interfere with another life-form is disrespectful and another form of cultural arrogance."

A more practical objection to transgenic technology is the risk of altering **ecosystems**. Consider genetically modified Atlantic salmon, currently under review at the U.S. Food and Drug Administration (FDA). The fish have a growth hormone gene taken from Chinook salmon and a DNA sequence that controls the gene's expression taken from ocean pout, a fish that produces the hormone year-round. Because Atlantic salmon normally produce growth hormone only during the summer, the transgenic animal grows at more than twice the natural rate. Such genetically modified salmon could escape the farms where they are intended to be raised and invade natural ecosystems, where they may outcompete native fish for space, food, and mates.

Until recently, the fear that a transgenic organism might escape and infiltrate a natural ecosystem was based on theoretical scenarios. For example, a 1999 report of transgenic corn pollen harming Monarch butterfly larvae in a laboratory simulation was not confirmed by larger, more realistic studies. But in 2001 transgenic corn was discovered growing on remote mountaintops in Mexico, ironically in the area where most natural corn variants originated. The corn was not supposed to have been able to spread beyond the fields where it was grown. At about the same time, 10,000 hectares (24,700 acres) of transgenic cotton were found in India. A farmer had crossed transgenic cotton he had obtained from the United States with a local variant and planted crops, not realizing that he had used a genetically modified product.

At the present time, American consumers cannot tell whether a food contains a genetically modified product or not because the two-thirds of processed foods that include GMOs and are sold in the United States have not been labeled. This lack of labeling is consistent with existing regulatory practice. While the FDA tests foods to determine their effect on the human digestive system, their biochemical makeup, and their similarity to existing foods (using a guiding principle called substantial equivalence), foods are not judged solely by their origin. For example, the FDA denied marketing of a potato derived from traditional selective breeding

Greenpeace activists hang a banner on Kellogg's "Cereal City," a company museum in Battlecreek, Michigan. Greenpeace asked Kellogg's to stop what Greenpeace perceived as Kellogg's double standard regarding genetically modified ingredients in March 2000.

that produces a toxin, while allowing marketing of a transgenic potato that has a high starch level and therefore absorbs less cooking oil, and is nontoxic. The FDA and U.S. Department of Agriculture approved transgenic crops in 1994, and deregulated the technology two years later, as did the U.S. Environmental Protection Agency. Ironically, as people in wealthier nations object to not having a choice in avoiding genetically modified foods, others complain that the technology is too expensive for farmers in developing nations to use.

Another ethical dimension to transgenic organisms is that the methods to create genetically modified seeds, and the seeds themselves, lie in the hands of a few multinational corporations. In the mid-1990s a

company sold transgenic plants resistant to the company's herbicides, but that could not produce their own seed, forcing the farmer to buy new seed each year. An international outcry led to the abandonment of this practice, but the use of crops that are resistant to certain herbicides, with a single company owning both seed and herbicide, continues. Some see this as a conflict of interest.

Groups that oppose genetically modified foods sometimes behave unethically. In 1999 environmental activists destroyed an experimental forest of poplars near London. The trees were indeed transgenic, but the experiments were designed to see if the trees would require fewer chemical herbicides, an activity the environmentalists had themselves suggested. More alarming were several incidents in the United States in 2000, when people who object to genetically modified foods vandalized laboratories and destroyed fields of crops, some of which were not even transgenic.

So far, foods containing GMOs appear to be safe. They may be easier to cultivate and may permit the development of new variants. However, it will take more time to determine whether or not they have longer term health and ecological effects. SEE ALSO BIOTECHNOLOGY: ETHICAL ISSUES; GENETIC TESTING: ETHICAL ISSUES; GENETICALLY MODIFIED FOODS; TRANSGENIC ANIMALS; TRANSGENIC PLANTS.

Ricki Lewis

Bibliography

Charles, Dan. *Lords of the Harvest.* Cambridge, MA: Perseus Publishing, 2001.

Dalton, Rex. "Transgenic Corn Found Growing in Mexico." *Nature* 413 (September 27, 2001): 337.

Yoon, C. K. "Altered Salmon Leading Way to Dinner Plates, but Rules Lag." *New York Times* (May 1, 2000): A1.

Transgenic Plants

Transgenic plants are plants that have been genetically modified by inserting genes directly into a single plant cell. Transgenic crop plants modified for improved flavor, pest resistance, or some other useful property are being used increasingly.

Transgenic plants are unique in that they develop from only one plant cell. In normal sexual reproduction, plant offspring are created when a pollen cell and an ovule fuse. In a similar laboratory procedure, two plant cells that have had their cell walls removed can be fused to create an offspring.

Genetic Engineering Techniques

There are three general approaches that can be used to insert the DNA into a plant cell: **vector**-mediated transformation, particle-mediated transformation, and direct DNA absorption. With vector-mediated transformation, a plant cell is infected with a virus or bacterium that, as part of the infection process, inserts the DNA. The most commonly used vector is the crown-gall bacterium, *Agrobacterium tumefaciens*. With particle-mediated transformation (particle bombardment), using a tool referred to as a "gene gun," the DNA is carried into the cell by metal particles that have been acceler-

vector carrier

A "normal" melon (left) after five days, and a transgenic melon (right) after fifteen days. The ability of the modified fruit to be edible for a much longer period is a trait desirable for those in the produce business.

ated, or "shot," into the cell. The particles are usually very fine gold pellets onto which the DNA has been stuck. With direct DNA absorption, a cell is bathed in the DNA, and an electric shock usually is applied ("electroporation") to the cell to stimulate DNA uptake.

No matter what gene insertion method is used, a series of events must occur to allow a whole genetically modified plant to be recovered from the genetically modified cell: The cell must incorporate the new DNA into its own chromosomes, the transformed cell must initiate division, the new cells need to organize themselves into all the tissues and organs of a normal plant ("regeneration"), and finally, the inserted gene must continue to work properly ("**gene expression**") in the regenerated plant.

gene expression use of a gene to create the corresponding protein

To help ensure all this occurs, a "cassette" of genes is inserted during the initial transformation. In addition to the gene coding for the desired trait, other genes are added. Some of these genes promote the growth of only those plant cells that have successfully incorporated the inserted DNA. It might do this by providing the transformed cells with resistance to a normally toxic antibiotic that is added to the growth medium, for example. Other genes ("**promoters**") may be added to control the functioning of the trait gene by directing when and where in the transformed plant it will operate.

promoters DNA sequences to which RNA polymerase binds to begin transcription

The genes put into plants using genetic engineering can come from any organism. Most genes used in the genetic engineering of plants have come from bacteria. However, as scientists learn more about the genetic makeup of plants ("plant genomics"), more plant-derived genes will be used.

Agricultural Applications

Inserted genes can be classified into three groups based on their use: those that protect a crop, those that improve the quality of a harvested product, and those that let the plant perform some new function.

Genes That Protect a Crop. The major use of plant genetic engineering has been to make crops easier to grow by decreasing the impact of pests. Insect resistance has been achieved by transforming a crop using a Bt gene. Bt genes were isolated from *Bacillus thuringiensis*, a common soil bacterium. They code for proteins that severely disrupt the digestive system of insects. Thus an insect eating the leaf of a plant expressing a Bt gene stops eating and dies of starvation. There are many Bt genes, each of which targets a particular group of insects. Some Bt genes, for example, target caterpillars. Others target beetles.

Genetic engineering also has been used in the battle against weeds. Bacterial genes allow crops to either degrade herbicides or be resistant to them. The herbicides that are used are generally very effective, killing most plants. They are considered environmentally benign, degrading rapidly in the soil and having little impact on humans or other organisms. Thus a whole field of transgenic crops can be sprayed with broad-spectrum herbicides, killing all plants except the crops. Corn, soybeans, canola, and cotton that have been engineered to withstand either insects or herbicides, or both, are widely planted in some countries, including the United States. In addition, other crops, including potatoes, tomatoes, tropical fruits, and melons, have been engineered for resistance to viral diseases.

Genes That Improve Crop Quality. An emerging major use of genetic engineering for crops is to alter the quality of the crop. Fresh fruits and vegetables begin to deteriorate immediately after being harvested. Delaying or preventing this deterioration not only preserves a produce's flavor, and appearance, but maintains the nutritional value of the produce. Genes that change the hormonal status of the harvested crops are the major targets for genetic engineering toward longer shelf-life.

For example, the plant hormone ethylene is associated with accelerated ripening, as well as leaf and flower deterioration, in fruits that are injured or harvested. Scientists insert genes that interfere with a plant's ability to synthesize or respond to ethylene, thereby extending postharvest quality for many fresh products, including tomatoes, lettuce, and cut flowers. Scientists are also using gene insertion to improve a plant's nutritional value and color.

Genes That Introduce New Traits. One approach to improving the economic value of crops is to give them traits that are completely new for that plant. Some crops, including potatoes, tomatoes, and bananas, have been engineered with genes from **pathogenic** organisms. This is done to make animals, including humans, that eat the crops immune to the diseases caused by the pathogens. The genes code for proteins that act as **antigens** to induce immunity. Thus edible parts of plants are engineered to act as oral vaccines. This approach may be particularly effective for pathogens, such as those causing diarrhea and other gastrointestinal disorders, that enter the body through **mucous membranes**. This is because the "medicine" in the food comes into direct contact with these membranes and does not have to be absorbed into the blood stream. Genes have also been engineered into crop

pathogenic disease-causing

antigens foreign substances that provoke an immune response

mucous membranes nasal passages, gut lining, and other moist surfaces lining the body

plants to direct the plants to produce industrial enzymes used in the manufacture of paper. Other genes direct crops to produce small polymers useful in the manufacture of plastics. This general approach is being termed "plant molecular farming."

Rice is another plant that has been engineered for a new trait. During commercial processing, a substantial part of the white rice grains are removed, leaving very little vitamin A. Vitamin A deficiency is a significant health problem in regions dependent on rice as a dietary staple. Scientists engineered a certain form of rice, known as "golden rice" because it has a yellow tinge, to express three introduced genes. These genes let the plant produce the **precursor** of vitamin A in the portion of the grain that remains after processing, thereby providing a dietary source of the vitamin. SEE ALSO AGRICULTURAL BIOTECHNOLOGY; BIOPESTICIDES; BIOTECHNOLOGY; CLONING GENES; CLONING ORGANISMS; GENE TARGETING; GENETICALLY MODIFIED FOODS; PLANT GENETIC ENGINEER; TRANSFORMATION.

Brent McCown

precursor a substance from which another is made

Translation

Translation is the cellular process in which the genetic information carried by the DNA is decoded, using an RNA intermediate, into proteins. This process is also known as protein synthesis.

Deciphering the Genetic Code

There are two steps in the path from genes to proteins. In the first step, called **transcription**, the region of the double-stranded DNA corresponding to a specific gene is copied into an RNA molecule, called messenger RNA (mRNA), by an enzyme called RNA polymerase. In the second step, called translation, the mRNA directs the assembly of amino acids in a specific sequence to form a chain of amino acids called a **polypeptide**. This process is accomplished by **ribosomes**, special amino acid–bearing RNA molecules called transfer RNAs (tRNAs), and other translation factors. The newly synthesized polypeptides form proteins, which have functional and structural roles in cells. All proteins are synthesized by this process.

The precise order of amino acids assembled during translation is determined by the order of **nucleotides** in the mRNA. These nucleotides are a direct copy of the linear sequence of the nucleotides in one of the two complementary DNA strands, which have been transcribed using a code in which every three bases of the RNA specify an amino acid. DNA and RNA molecules both have directionality, which is indicated by reference to either the 5′ ("five prime") end or the 3′ ("three prime") end.

The code is always read in the 5′ to 3′ direction, using adjacent, non-overlapping three-base units called codons. Since there are four different nucleotides (also called bases) in RNA (abbreviated A, C, G, and U), there are sixty-four (4^3) different codons, and each codon specifies a particular amino acid. There are only twenty different amino acids, however, so many of the amino acids can be specified by more than one codon, a circumstance that is known as degeneracy. The list of mRNA codons specific for a given

transcription messenger RNA formation from a DNA sequence

polypeptide chain of amino acids

ribosomes protein-RNA complexes at which protein synthesis occurs

nucleotides the building blocks of RNA or DNA

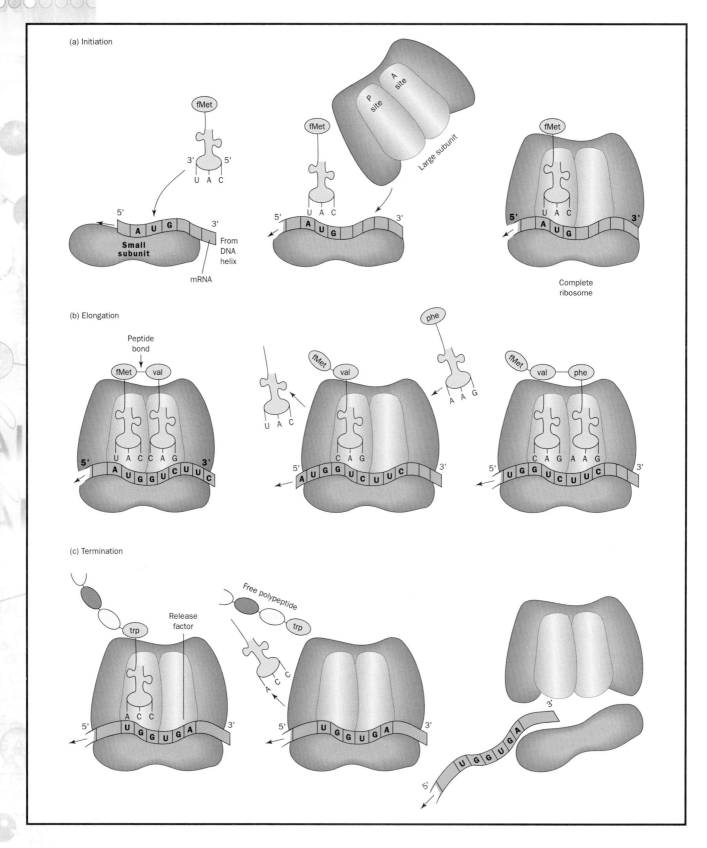

The steps of translation. Initiation: RNA binds to the small ribosomal subunit, a tRNA binds to the RNA, and the large subunit attaches to the small subunit. Elongation: Successive tRNAs bind to the A site, form a peptide bond to the growing amino acid chain, and then move to the P site. The spent tRNA is ejected. Termination: A termination or stop codon binds a release factor, which dissociates two ribosomal subunits. The peptide (amino acid chain) is released. Adapted from Curtis and Barnes, 1994.

Genetic code.

		Second base			
		U	C	A	G
First base	U	UUU } Phe UUC } UUA } Leu UUG }	UCU } UCC } Ser UCA } UCG }	UAU } Tyr UAC } UAA } STOP UAG }	UGU } Cys UGC } UGA } STOP UGG } Trp
	C	CUU } CUC } Leu CUA } CUG }	CCU } CCC } Pro CCA } CCG }	CAU } His CAC } CAA } Gln CAG }	CGU } CGC } Arg CGA } CGG }
	A	AUU } AUC } Ile AUA } AUG } Met	ACU } ACC } Thr ACA } ACG }	AAU } Asn AAC } AAA } Lys AAG }	AGU } Ser AGC } AGA } Arg AGG }
	G	GUU } GUC } Val GUA } GUG }	GCU } GCC } Ala GCA } GCG }	GAU } Asp GAC } GAA } Glu GAG }	GGU } GGC } Gly GGA } GGG }

amino acid is called the **genetic code**. The start signal, or initiation codon, for translating the mRNA is usually specified by an AUG, which codes for the amino acid methionine. Three codons (UAA, UGA, and UAG) do not specify an amino acid. Instead, these codons serve as stop signals to indicate that the end of the gene has been reached. During the translation process, they signal that no further amino acids are to be assembled.

genetic code the relationship between RNA nucleotide triplets and the amino acids

The process of translation is carried out by ribosomes, which bind the mRNA and conduct a **catalytic** activity, called **peptide bond** formation, for joining the amino acids. The amino acids are carried to the ribosome by the tRNAs. Each tRNA has a specific amino acid attached to it and contains a nucleotide triplet called an anticodon. The anticodon recognizes a specific codon on the mRNA by pairing with it, using base-pairing rules like those used by DNA: A pairs with U and G pairs with C. For example, a tRNA with a UUU anticodon recognizes the AAA codon. The amino acid lysine is attached to this tRNA, so every time the ribosome "reads" an AAA codon, the lysine-bearing tRNA is brought in, base pairs via its anticodon to the codon, and delivers a lysine to the growing protein chain.

catalytic describes substances that speed up a reaction without being consumed (e.g., enzyme)

peptide bond bond between two amino acids

Mutations arise when one or more bases in the DNA is changed. When the mutated DNA is transcribed, the resulting mRNA will carry the same mutation. Then, when the mRNA is translated, the amino acid sequence of the resulting protein will be different from the original, or **wild-type**, sequence because the codons affected by the mutation will recruit the wrong amino acids. The resulting mutant protein may have neutral, harmful, or even beneficial effects on the individual. These changes are the basis for evolution.

wild-type most common form of a trait in a population

Stages of Translation

The process of translation can be broken down into three stages. The first stage is initiation. In this step, a special "initiator" tRNA carrying the amino acid methionine binds to a special site on the small subunit of the ribosome (the ribosome is composed of two subunits, the small subunit and the large

subunit). The mRNA is also loaded on, and positioned so that the initiation codon (usually AUG) is base paired with the anticodon of the initiator tRNA. The large subunit then binds to the small subunit. The resulting complex of ribosome, mRNA, and methionine-bearing initiator tRNA is called an initiation complex. Formation of this complex also requires a number of helper proteins called initiation factors.

The second stage is called chain elongation. During this stage, additional amino acids are progressively added. The methionine-bearing initiator tRNA sits on a site of the ribosome called the P (peptidyl) site. A new tRNA, bearing the next amino acid is base paired via its anticodon to the next codon of the mRNA, using a site called the A (acceptor) site. This new amino acid is then attached to the amino acid carried by the P site tRNA, forming a peptide bond. This enzymatic step is carried out by the ribosome, at a site called the peptidyl-transferase center.

The tRNA that has so far been attached to the amino acid in the P site is then released through the E (exit) site, and the new tRNA, now carrying both its own amino acid and the methionine moves into the P site. The mRNA also slides three bases to bring the next codon into position at the A site. A third tRNA, again carrying a specific amino acid and recognizing the third codon of the mRNA, moves into the A site, and the cycle is repeated. As these steps are continued, the mRNA slides along the ribosome, three bases at a time, and the peptide (amino acid) chain continues to grow. As with initiation, elongation requires helper proteins, called elongation factors. Energy is also required for peptide bond formation.

The final stage of translation is termination. The signal to stop adding amino acids to the polypeptide is a stop codon (UAA, UAG, or UGA), for which there is no partner tRNA. Rather, special proteins called release factors bind to the A site of the ribosome and trigger an enzymatic reaction by the ribosome. This reaction causes the ribosome to release the polypeptide and mRNA, ending the elongation process.

At a given time, more than one ribosome may be translating a single mRNA molecule. The resulting clusters of ribosomes, which resemble beads on a string, are called polysomes.

Recognition of Initiation Codons

Not all AUG codons serve as the site of initiation. Most AUGs are intended to code for methionines within the polypeptide chain. Therefore, in addition to the methionine-bearing initiator tRNA, another set of methionine-specific tRNAs are used for these internal AUG codons. The ribosome must be able to distinguish between these two kinds of AUG codons. In bacteria, additional information contained within the mRNA sequence immediately before the intended initiating AUG, called a Shine-Dalgarno sequence, helps the ribosome to recognize where it should start translating. Any AUG sequences on the 5′ side of the initiation codon are ignored. In eukaryotic cells, a different strategy is used to recognize the initiating AUG codon. The mRNA contains a special structure at its 5′ end, which helps the ribosome to attach and then to scan down the RNA molecule until it reaches the first AUG triplet. In bacteria and eukaryotes, AUG codons encountered during translation after initiation are recognized by a non-initiator methionine-bearing tRNA. SEE ALSO GENETIC CODE; MUTATION; NUCLEOTIDE; PROTEINS; READING FRAME; RIBOSOME; RNA; TRANSCRIPTION.

Janice Zengel

Bibliography

Lewin, Benjamin. *Genes VII*. New York: Oxford University Press, 2000.

Transplantation

Modern medicine continues to offer many miracles that lengthen the life spans of humans, as well as greatly increase the quality of life that they enjoy. If one were to draw up a "top ten" list of technical feats, surely the ability to successfully transplant an entire organ from one human to another would be high on the list. Transplantation can be defined as the transfer of cells, tissues, or organs from one site in an individual to another, or between two individuals. In the latter case, the individual who provides the transplant organ is termed a donor, and the individual receiving the transplant is known as the recipient. The science of transplant biology has, in fact, become a victim of its own success, in that the demand for organs exceeds the supply of donors.

Types of Transplants

There are four basic types of transplants, which reflect the genetic relationship of the recipient to the donor. The autograft is the transfer of tissue from one location of an individual's body to another location that is in need of healthy tissue; in other words, the recipient is also the donor. Common examples of autografts are skin transplants in burn patients and bypass surgery in patients suffering from coronary heart disease. The syngraft is a transplantation procedure carried out between two genetically identical individuals. These types of transplants, like autografts, are always successful, unless there have been technical problems during the surgery. The first successful human kidney transplant was a syngraft, carried out in 1954 between identical twins.

An allograft is the transfer of tissue or an organ between nonidentical members of the same species. This is the predominant form of transplantation today, and allografts have dominated transplant research for many years. Finally, the xenograft represents the most disparate of genetic relationships, because it is the transfer of tissue or organs between members of different species. Many think that xenografts are the answer for solving the shortage of transplant tissue and organs that we are currently experiencing. Both allografts and xenografts have the disadvantage that the recipient's immune system is designed to recognize and reject foreign tissue.

The Genetic Basis of Transplant Rejection

Research that began in the 1940s gave geneticists the first hints that a portion of the mammalian **genome** contained a cassette of genes that governed the acceptance or rejection of transplanted tissues. This grouping of genes was labeled the major histocompatibility complex (MHC). Subsequently, it has been found that the MHC also contains genes that are

genome the total genetic material in a cell or organism

Five "knockout" piglets, considered to be the first of their kind, were raised with the intention of using their organs for human transplants. Knockout pigs have the gene inactivated that usually leads human immune systems to rejecting pig organs.

antigen a foreign substance that provokes an immune response

involved in governing antibody responses as well. MHC molecules are identical between identical twins, but are otherwise different for every individual. Thus they allow the body to distinguish "self" from "nonself" on the molecular level.

The immunogenicity (ability to induce an immune response) of major transplantation **antigens** is so strong that differences between the antigens of the donor and recipient is enough to trigger an acute rejection response. To the extent that it is possible, therefore, the recipient and donor are matched for MHC type, to minimize acute rejection.

However, there are cases in which the donor and recipient are very well matched, and yet rejection of the graft still occurs. This is due to other genes found in various places in the genome, known as minor histocompatibility genes, that encode for other weaker transplantation antigens, or foreign peptides, that can cause a chronic rejection response. Currently, researchers have not been able to determine the extent or location of all of these genes. Results obtained from the mapping of genes in the human genome will aid in overcoming this problem.

The Mechanisms of Transplant Rejection

The immune system's attack on foreign tissue is mediated by lymphocytes, phagocytic cells, and various other white blood cells. Various subgroups of lymphocytes have different responsibilities. Once stimulated, the B-lymphocytes (derived from bone marrow) will develop into a cell that produces antibodies (soluble proteins that specifically seek out invaders). Antibodies may cause hemorrhaging by attaching to the lining of blood vessels in the transplant and then activating a naturally occurring series of potent enzymes known as the complement system.

The T-lymphocyte (derived from the thymus) can develop either into a T-helper cell, which serves a regulatory function, or a T-cytotoxic (killer) cell. Activated T-helper cells induce T-cytotoxic cells to destroy a foreign graft by attacking those cells in the transplant that display incompatible antigens.

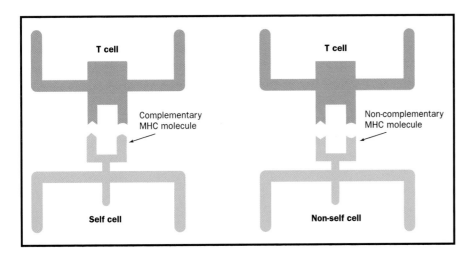

When a T cell detects a noncomplementary MHC molecule on a cell, it interprets it as "nonself" and sets a rejection reaction in motion.

The task of the transplant treatment team is to somehow derail this natural process of reacting to foreign tissue long enough for the graft to "heal in" and survive without at the same time putting the patient at risk for increased infectious disease. To control this type of response, various immunosuppressive drugs, such as cyclosporine, have been developed. Great strides have been made in controlling rejection of transplated tissue and organs by these methods.

The Supply Crisis in Transplantation

The predominant issue in transplantation biology is now one of increasing the supply of organs for patients in need of them. This is not only a technical problem, but in some cases, raises ethical issues as well. For instance, there have been cases of parents with a sick child purposely conceiving a second child for the main purpose of being a bone marrow donor for their ailing offspring. There has also been the rise of a black market in body parts, particularly emanating from China, in which various organs from executed prisoners are offered for sale.

Researchers have come up with numerous new options to improve on the availability of organs needed for transplantaion. For instance, chemicals can be used to stimulate a patient's own stem cells (cells that can develop into almost any type of tissue, depending upon the local influences it encounters) to migrate from the bone marrow to the diseased organ, develop into the right type of cell, and regenerate the organ. A more controversial application of stem cell research involves the use of embryonic stem cells. One version of this strategy is to remove DNA from the patient's own skin cells, inject it into a donated human egg from which the nucleus has been removed, and then allow that egg to develop into an early-stage embryo. The embryo can then be harvested for embronic stem cells that can be influenced into growing into the organ of choice. Another major strategy is to collect embryonic stem cells from aborted fetuses or from umbilical cord blood. This whole topic has become a very highly debated issue due to the involvement of human embryos, as has the entire burgeoning field of stem cell–applied medical treatment.

Adapted from Roitt, 2001.

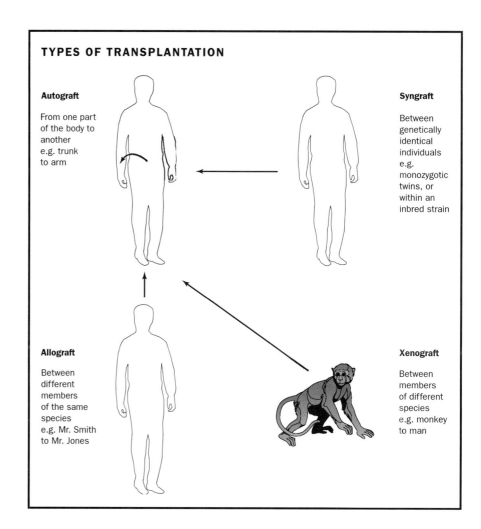

TYPES OF TRANSPLANTATION

Autograft

From one part of the body to another e.g. trunk to arm

Syngraft

Between genetically identical individuals e.g. monozygotic twins, or within an inbred strain

Allograft

Between different members of the same species e.g. Mr. Smith to Mr. Jones

Xenograft

Between members of different species e.g. monkey to man

Another option is called tissue scaffolding. The strategy in this technique is to take the patient's cells (or cells from a donor) and inject them into a three-dimensional scaffold of biodegradable polymers that are in the shape of the desired organ. The entire structure is transplanted, and the cells in the scaffold replicate, reorganize, and form a new organ. Research indicates that cells are surprisingly adept at regenerating the tissue of their origin. As the cells grow, and the polymers of the scaffold naturally degrade, what is left is a new, functioning organ. The obvious problem with this technique, as with options using stem cells, is that it takes time to form the new organ. This is of little help to a patient who is in immediate need of a transplant.

One of the most promising, and controversial, sources for new organs for humans are xenotransplants from other species, particularly baboons and pigs. Many individuals are very strongly opposed to raising animals for the sole purpose of harvesting their organs for humans, viewing it as inhumane. Another area of controversy, particularly concerning baboon donors, is the possibility of spreading unknown diseases into the human population. There are already established precedents for viral diseases jumping from primates to humans, such as the AIDS virus (HIV), Ebola virus, and the hantavirus. Consequently, there is a fear that xenotransplantaion could unleash a new plague upon humans. More and more xenotransplant research is moving toward the use of pigs, since it is very much less likely that a pig virus could infect a human. The development of **pathogen**-free colonies of pigs would also greatly reduce the likelihood of such an occurrence.

The real advantage to using pigs is that they are easily bred, mature quickly, and their organs are of a comparable size to that of humans. In addition, pigs are amenable to genetic engineering, whereby the genes that encode transplantation antigens that would be recognized by a human recipient could be removed so that the resulting organs would not be recognized as foreign in the human. In addition, pigs have now been cloned, so that once such an antigen-free animal has been constructed, we could have a continuous source of immunologically nonstimulating organs available for transplantation into human patients. SEE ALSO AGRICULTURAL BIOTECHNOLOGY; CLONING ORGANISMS; EMBRYONIC STEM CELLS; IMMUNE SYSTEM GENETICS.

Richard D. Karp

Bibliography

Colen, B. D. "Organ Concert." *Time Magazine* (Fall 1996): 70–74.

Goldsby, R. A., T. J. Kindt, and B. A. Osborne. *Kuby Immunology*, 4th ed. New York: W. H. Freeman, 2000.

Lanza, R. P., D. K. Cooper, and W. L. Chick. "Xenotransplantation." *Scientific American* 277, no. 7 (1997): 54–59.

Miklos, A. G., and D. J. Mooney. "Growing New Organs." *Scientific American* 280, no. 4 (1999): 60–65.

Roitt, Ivan M., Jonathan Brostoff, and David K. Male. *Immunology*. St. Louis: Mosby, 2001.

Transposable Genetic Elements

Transposable genetic elements (TEs) are segments of DNA that can be integrated into new chromosomal (genomic) locations either through direct DNA transfer (transposons), or via an RNA intermediate (retrotransposons). Pseudonyms for TEs include mobile elements, jumping genes, genomic parasites, and selfish DNA. TEs are known to be responsible for several human genetic diseases and may play a role in evolution in many species.

Early Evidence

Barbara McClintock originally theorized that unusual patterns of phenotypic variance (in corn kernels) could be explained by gene transposition. However, this explanation did not coincide with traditional Mendelian inheritance that genetic information was fixed within the **genome**, and her views were not widely accepted within the scientific community until the 1960s, when evidence of transposition began to accumulate.

McClintock was a cytogeneticist working on maize. She noted that, in maize, there was a pigment-bearing layer of the kernel, called the aleurone layer, that changed color from kernel to kernel and generation to generation. She also noted a baffling result that occurred when a **homozygous** plant for purple aleurone (CC) was crossed with a colorless aleurone homozygote (cc). About one-half of the kernels of offspring corn were solid purple, and one-half were purple with varying sizes of colorless spots, suggesting breakage (and loss) of the C locus. However, one of 4,000 analyzed

pathogen disease-causing organism

genome the total genetic material in a cell or organism

homozygous containing two identical copies of a particular gene

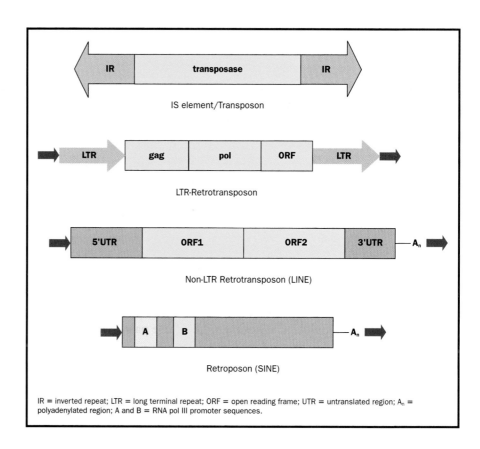

Figure 1. Schematic diagrams of various classes of transposable elements.

IR = inverted repeat; LTR = long terminal repeat; ORF = open reading frame; UTR = untranslated region; A_n = polyadenylated region; A and B = RNA pol III promoter sequences.

locus site on a chromosome (plural, loci)

kernels was colorless with purple spots, indicating a "gain-of-function" of the c locus. McClintock identified a **locus** (called *Ds*, which stands for dissociation) that induced nearby chromosome breakage only in the presence of another gene, which she called *Ac* (for activator). She concluded that the inactivation of pigment production is caused by insertion of the *Ds* gene into the C locus (disrupting pigment products and yielding the colorless background), and the eventual movement of the *Ds* out of the gene, restoring pigment production and yielding the purple spots. All of this could occur within a single kernel.

Further support for the mobility of the *Ds* and *Ac* genes was the inability to determine their chromosomal locations, which differed among plants. High mutational reversion rates have been subsequently identified in other organisms, with mobile elements now offering a plausible explanation.

TEs across the Evolutionary Tree

TEs are ubiquitous throughout the evolutionary tree from microorganisms to mammals. For instance, a bacterial virus called phage MuM (mutator) is characterized as a TE based on similarities with other transposons. Mu integration into the host bacterial chromosome is considered transposition because it can occur nearly anywhere, thereby inactivating host genes and generating insertions and deletions.

Transposons also occur naturally in bacterial genomes. The extensively studied species *Escherichia coli* (*E. coli*) contains insertion sequences (IS) and transposons in its genome. IS elements are generally small, have inverted sequence repeats at their ends (important for their mobility), and contain

Transposase is an enzyme that catalyzes transposition. It may be encoded in the transposon or may reside elsewhere.

an overlapping genetic region encoding a transposase and a repressor. Transposons are larger than IS elements, since they contain additional genes such as drug-resistance genes. These elements are flanked either by inverted repeats or by IS elements. Bacterial transposons undergo conservative transposition, in which the transposon is excised and pasted elsewhere, or replicative transposition, in which it is copied and the copy is inserted elsewhere.

Ty elements in yeast contain retroviral-related sequences (called gag and pol) and include long terminal repeats (LTRs); hence they are considered viral (or viral-like) retrotransposons or LTR retrotransposons. Their activity is replicative: An RNA is transcribed from the gene, reverse-transcribed to DNA (cDNA) and integrated elsewhere in the genome. Since Ty does not contain *env* genes, which code for encapsulating envelope proteins, it does not yield infectious particles. However, viruslike particles accumulate in cells in which retroposition has been induced.

The African Trypanosome, a parasitic protozoan, contains the ingi (non-LTR) retrotransposons. Ingi may therefore be referred to as a parasite's parasite. Full-length elements are 5.2 kilobase pairs long, have multiple adenine nucleotides at one end (called a "poly A tail"), and DNA sequences similar to reverse transcriptase genes and mammalian LINEs (discussed below). Among insects, the *Drosophila* genome contains a virtual cornucopia of TEs. Fruit fly transposons include mariner, hobo, and *P* elements, non-LTR-retrotransposons include *I*, *F*, and *jockey* elements, and LTR-retrotransposons such as gypsyM and copia-like elements. These eukaryotic transposons are similar to bacterial IS elements, but are generally larger due to the presence of introns (noncoding sequences of the genome).

The primary TEs in mammalian genomes include short and long interspersed repetitive elements (SINEs and LINEs, respectively). SINEs represent a group of small retrotransposons (75–500 **base pairs**) and lack protein-coding sequences. In primates, *Alu* elements represent the predominant SINE family. More than 1 million copies of Alu are contained in the human genome, representing about 13 percent of the genome. This is truly impressive considering their lack of replicative autonomy. Alu elements are 300 base pairs in length, are rich in adenine sequences both internally and at the 3′ (downstream) end, and internal RNA polymerase III promoter sequences, allowing them to be replicated. L1 elements represent the predominant primate LINE family, contain two **open reading frames**, adenine-rich 3′ ends, and internal RNA polymerase II promoter sequences. They constitute approximately 20 percent of the human genome. A full-length LINE is about 6.5 kilobase pairs long, although most elements are truncated as a result of incomplete reverse transcription.

Transmission of TEs

TEs generally demonstrate vertical transmission, meaning that new incorporations are inheritable by offspring. Although transposons may excise from their genomic location and integrate elsewhere, retrotransposons form stable integrations, creating a molecular "fossil record" of past integration events.

Over 99 percent of human Alu elements are shared with the chimpanzee genome. Unlike Alu, L1 predates the origin of primates, with many

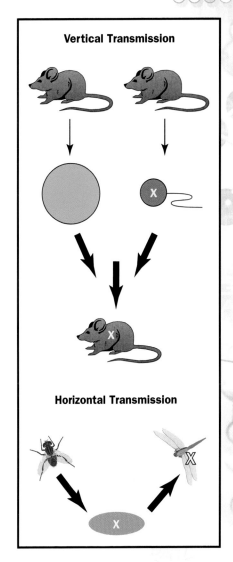

Vertical Transmission

Horizontal Transmission

Figure 2. Transmission of transposable elements (TEs). Vertical transmission involves a germ-line TE integration (indicated by the X in the sperm), that is passed on to the next generation. Horizontal transmission involves the transfer of a TE (indicated by the X) across species, possibly via a parasite.

base pairs two nucleotides (either DNA or RNA) linked by weak bonds

open reading frames DNA sequences that can be translated into mRNA; from start

Figure 3. Creation of presence/absence variants (dimorphism) by the integration of an Alu element. The Alu increases the gene size by 300 bp.

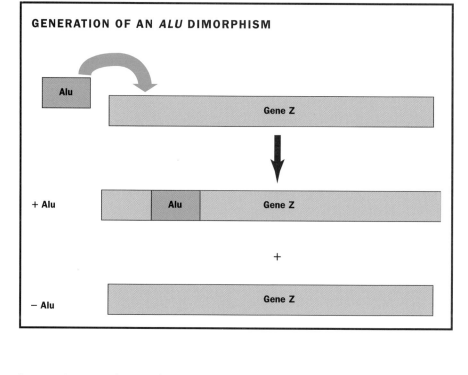

GENERATION OF AN *ALU* DIMORPHISM

Alu

Gene Z

+ Alu

Alu   Gene Z

+

− Alu

Gene Z

integrations predating the origin of placental mammals. Most recent Alu and L1 integrations in the human genome generate insertion presence/absence variants that are referred to as dimorphisms.

Dimorphisms provide DNA markers useful in mapping studies, fingerprinting, and human population investigations. Retrotransposon integrations are also associated with human disorders. Examples include Alu integrations into NF1, factor IX, and BRCA2 genes yielding neurofibromatosis, hemophilia, and breast cancer, respectively, and L1 integrations into the *c-myc*, APC, and factor VIII genes cause breast cancer, colon cancer, and hemophilia, respectively.

Horizontal transfer of TEs, or the transmission between species, has also been implicated. General evidence involves the lack of **phylogenetic** correspondence between the TEs and their host organisms. For example, outside of the *Drosophila melanogaster* species group, *jockey* has been detected only in the distantly related *D. funebris*, suggesting that it was transferred between the two. *P* elements have also exhibited horizontal transfer, possibly from *D. willistoni* to *D. melanogaster*, but more importantly, strains of *D. melanogaster* are lacking the element, indicating recent spreading through populations. Additionally, insect-related mariner elements, characterized primarily by their transposase, have been identified in diverse organisms such as flatworms and hydra, yet are lacking in twenty other invertebrate species representing major phyla. These transfers are both ancient and relatively recent, as one element has 92 percent amino acid similarity between Hydra and a staphylinid beetle. Possible transmission **vectors** include parasites (such as mites) and viruses. Some evidence also exists for horizontal transfer of mammalian TEs, including the putative discovery of an Alu element in the malarial parasite *Plasmodium vivax*.

phylogenetic related to the evolutionary development of a species

vectors carriers

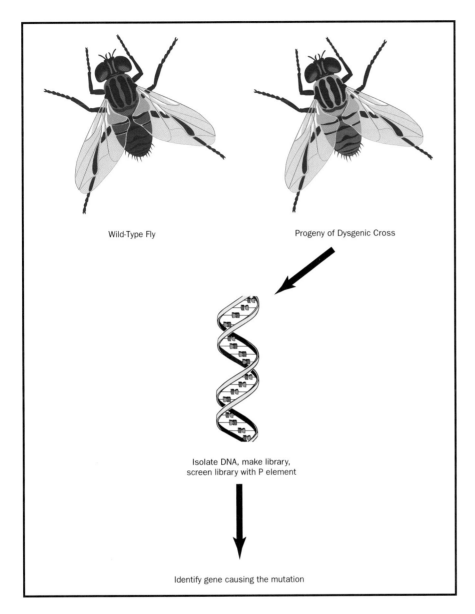

Figure 4. Identifying genes using transposable elements (TE). A dysengic fly with a randomly incorporated mutation is generated. DNA is isolated to produce a library which is screened with the TE to identify the mutated gene.

Wild-Type Fly

Progeny of Dysgenic Cross

Isolate DNA, make library, screen library with P element

Identify gene causing the mutation

Transposition Mechanisms

Transposon mobility may be either nonreplicative or replicative. In non-replicative mobility, the TE is cut out of its original position and integrated at a new location. In replicative mobility, transposition makes staggered cuts in donor and recipient sites, followed by a complex transfer, replication, and resealing operation.

SINE and LINE generation originates from "master genes": Only a few of the thousands to millions of copies are capable of serving as the source for new integrating elements. The proposed details of copy formation and integration differ between the two types of generation.

TEs and Species Evolution

Are TEs simply genomic parasites, or have they had a major impact on the evolution of species? Certainly, a TE integration could have an immediate

detrimental impact on its host. However, specific and cumulative integrations may provide a mechanism for speciation. *Drosophila P* elements have been implicated in generating a reproductive barrier between strains. When females devoid of active elements are crossed with males of a strain with active elements, the TEs run rampant in developing germ-line cells, yielding various chromosomal anomalies and F_1 hybrid sterility. This may be an important mechanism in promoting the creation of new species from those strains.

Arguably, SINEs and LINEs may also drive species evolution through several mechanisms. They can disrupt or reset coordinated gene regulation, facilitate the pairing of homologous chromosomes, and possibly offer sites for genomic imprinting. In addition to individual integrations, LINEs have contributed to genomic diversity by delivering adjacent sequences, including whole genes, to new genomic locations. TEs also offer numerous sites for homologous unequal recombination.

Transposons as Molecular Biology Tools

Transposons can be used to facilitate cloning of genes, identify regulatory elements, and produce transgenic organisms. For example, transposon tagging involves inducing transposition of a TE, allowing for disruption of a gene that generates an organism with a mutant phenotype, and is followed by molecular techniques that allow for the identification of the gene. A variation of transposon tagging (enhancer trapping) uses *P* elements to identify DNA sequences that regulate the expression of genes. *P* elements can also be used to incorporate foreign genes into fruit flies (transgenics). In addition, transposon fossils have been useful for the isolation of species-specific DNA from complex sources such as using inter-Alu PCR for the isolation of human genomic DNA sequences. SEE ALSO DNA LIBRARIES; EVOLUTION OF GENES; IMPRINTING; MCCLINTOCK, BARBARA; REPETITIVE DNA ELEMENTS; RETROVIRUS; REVERSE TRANSCRIPTASE; YEAST.

David H. Kass and Mark A. Batzer

Bibliography

Batzer, Mark A., and Prescott L. Deininger. "Alu Repeats and Human Disease." *Molecular Genetics and Metabolism* 67, no. 3 (1999): 183–193.

Griffiths, Anthony J. F., et al. *An Introduction to Genetic Analysis*, 7th ed. New York: W. H. Freeman, 2000.

Kass, David H. "Impact of SINEs and LINEs on the Mammalian Genome." *Current Genomics* 2 (2001): 199–219.

Lander, Eric S., et. al. "Initial Sequencing and Analysis of the Human Genome." *Nature* 409 (2001): 860–921.

Lewin, Benjamin. *Genes VII.* New York: Oxford University Press, 2000.

Lodish, Harvey, et al. *Molecular Cell Biology*, 4th ed. New York: W. H. Freeman, 2000.

Prak, Elaine T., and Haig H. Kazazian. "Mobile Elements and the Human Genome." *National Review: Genetics* 1 (2000): 134–144.

Watson, James D., et al. *Recombinant DNA*, 2nd ed. New York: W. H. Freeman, 1998.

Triplet Repeat Disease

Trinucleotide, or triplet repeats, consist of three consecutive nucleotides that are repeated within a region of DNA (for example, CCG CCG CCG

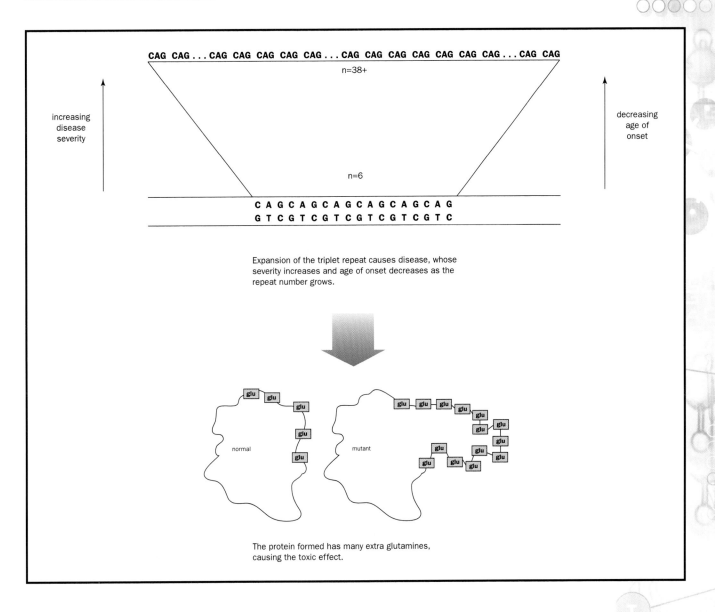

Expansion of the triplet repeat causes disease, whose severity increases and age of onset decreases as the repeat number grows.

The protein formed has many extra glutamines, causing the toxic effect.

CCG CCG). These are found in the **genome** of humans and many other species. All possible combinations of nucleotides are known to exist as triplet repeats, though some, including CGG and CAG, are more common than others. The repeats may be within genes or between genes. In genes, they may be in regions that specify proteins (coding regions called exons) or in noncoding regions (introns). If present within exons, they may be present in translated regions and hence encode a series of identical amino acids, or they may occur in regions not translated into protein. Triplet repeats are frequently found in genes that encode transcription factors and those involved in regulating development.

Implicated in Diseases

Triplet repeats were once thought to be apparently benign stretches of DNA. However, it is now known that these repeats can sometimes undergo dynamic or expansion mutation. In this type of mutation, through mechanisms during DNA replication that are only partly understood, the number of triplets in a repeat increases (expands) and can cause disease.

Expansion of a triplet repeat gene segment leads to increased disease severity and decreased age of onset in proportion to the degree of expansion. Repeats that are translated into protein alter the structure of the protein, conferring new functions.

genome the total genetic material in a cell or organism

somatic nonreproductive; not an egg or sperm

germ cells cells creating eggs or sperm

Diseases with Triplet Repeats in Noncoding Regions

Fragile X syndrome (CGG repeat)
Fragile XE syndrome (GCC repeat)
Friedreich ataxia (GAA repeat)
Myotonic dystrophy (CTG repeat)
Spinocerebellar ataxia type 8 (CTG repeat)
Spinocerebellar ataxia type 12 (CAG repeat)

Diseases with (CAG)$_n$ Repeats in Coding Regions

Spinobulbar muscular atrophy
  (Kennedy's disease)
Huntington's disease
Dentatorubral-pallidoluysian atrophy
Spinocerebellar ataxia types 1, 2, 3, 6 and 7

---

pathogenesis pathway leading to disease

phenotypes observable characteristics of an organism

kinase an enzyme that adds a phosphate group to another molecule, usually a protein

---

There are many diseases known to be caused by triplet repeats. They share certain common features. The mutant repeat length is unstable in both **somatic** and **germ cells** of the body, meaning it can change in length during DNA replication. Also, the triplet repeat often expands rather than contracts in successive generations. Increasing repeat size is correlated with decreasing age of onset or increasing disease severity in successive generations. This phenomenon is called anticipation, and is a characteristic of most triplet repeat diseases.

Classification of Triplet Repeat Disorders

Triplet repeat disorders fall into two subclasses, depending on the location of the trinucleotide repeat within the gene. The first subclass has triplet repeats occurring in noncoding sequences of DNA, either introns or regions at the start or end of the gene, called 5' ("five prime") or 3' ("three prime") untranslated regions. The second subclass is characterized by (CAG)$_n$ repeats that code for repeated stretches of the amino acid glutamine (polyglutamine) in coding regions of the affected gene.

Noncoding Triplet Repeat Disorders

The noncoding triplet repeat diseases typically have large and variable repeat expansions that result in multiple tissue dysfunction or degeneration. The triplet repeat sequences vary in this subclass (CGG, GCC, GAA, CTG, and CAG). It is clear that the particular triplet sequence and its location with respect to a gene are important defining factors in dictating the unique mechanism of **pathogenesis** for each disease. The pathogenic mechanism also varies from disease to disease depending on the consequences of the lost function of the respective proteins or in some cases acquired function of a toxic transcript.

Fragile X Syndrome. This disorder is caused by the expansion of a (CGG)$_n$ repeat in the 5' untranslated region of the fragile X mental retardation (*FMR1*) gene. Normally there are about six to fifty-three repeats in this region. In the disease state there are more than 230 triplet repeats. This causes the transcriptional silencing of the gene, leading to loss of normal gene function. Some of the symptoms seen in this disorder include mental retardation, deformed features, and hyperactivity.

Myotonic Dystrophy (DM). DM is a multisystem disorder with highly variable **phenotypes** and anticipation. Rigidity of muscles after contraction (tonic spasms), muscle weakness, and progressive muscle wasting characterize adult-onset DM. Developmental abnormalities, mental handicap, and respiratory distress are often evident in more severe forms. DM is caused by an expanded CTG triplet repeat in the 3' untranslated region of the protein **kinase** gene *DMPK*. The CTG expansion may disrupt *DMPK* transcription, causing loss of function. CTG expansions may also cause loss of function in two genes flanking *DMPK*: the DM locus-associated homeodomain protein (*DMAHP*) and gene 59 (also known as *DMWD*). The CTG expanded transcript could also gain toxic function by interfering with normal processing of various RNAs. Congenital myotonic dystrophy occurs from birth in infants whose mothers have DM, often a case so mild it is never diagnosed. Due to anticipation, the child is much more severely affected than the mother.

Polyglutamine Diseases

In comparison with the noncoding disorders, polyglutamine diseases have triplet (CAG) repeat expansions in coding regions of genes. Although the mutant proteins do not share any homology (similar sequences) outside the polyglutamine tract, the polyglutamine diseases have several similar features and may share common mechanisms of pathogenesis.

A simple loss of normal function of the gene does not account for the phenotype seen in these diseases. Studies of animal models and tissue culture systems have demonstrated that the mutant protein is toxic. The mutant protein can aggregate and form inclusions in the cytoplasm and nucleus. Also, all the polyglutamine diseases known so far are characterized by progressive neuronal dysfunction that typically begins in midlife and results in severe neurodegeneration. Despite the ubiquitous expression of the mutant genes, only a specific subset of neurons is vulnerable to neurodegeneration in each disease. What makes these specific neurons more susceptible than other neurons and other cells of the body remains a mystery.

Huntington's disease is caused by a CAG expansion in the *HD* gene that codes for a protein called huntingtin. This protein has no known function. In the normal, nonmutant form of the gene there are about six to thirty-five CAG repeats. In the disease-causing **allele** there are usually more than thirty-seven CAG repeats. Patients typically show symptoms of dementia, involuntary movements, and abnormal posture.

allele a particular form of a gene

Spinobulbar muscular atrophy (Kennedy's disease) is caused by CAG expansion in the *AR* gene that codes for the **androgen** receptor protein. The normal gene contains nine to thirty-six CAG repeats and the mutant gene usually contains thirty-eight to sixty-two repeats. This disease of motor neurons (nerve cells controlling muscle movement) is characterized by progressive muscle weakness and atrophy. Over 50 percent of the affected males may also have reduced fertility.

androgen testosterone or other masculinizing hormone

Why Are Polyglutamines Toxic?

Several hypotheses have been offered to explain the toxic nature of polyglutamine repeats, though none has yet been conclusively proven, and may yet be wrong in either details or central concept. The Nobel Prize recipient Max Perutz suggested that the expanded polyglutamine repeats promote the formation of protein aggregates. These aggregates often contain ubiquitin, a marker for protein degradation. Expanded polyglutamine proteins may adopt energetically stable structures that resist unfolding and therefore impede clearance by the cell's protein-degrading machinery, called the proteasome. This concept is supported by the observation that addition of proteasome inhibitors promotes aggregation of mutant huntingtin, ataxin-1, and ataxin-3 in cell culture.

Most polyglutamine aggregates are found in the nucleus. Is nuclear localization of the mutant proteins a critical event in polyglutamine disease, and if so, how do the mutant proteins affect nuclear function? Aggregates may interfere with such important events occurring in the nucleus as gene transcription, RNA processing, and nuclear protein turnover. The aggregates formed may sequester and deplete critical nuclear factors

required for transcription. Recent evidence suggests that mutant huntingtin does interfere with transcription factors.

Scientists have only begun to understand the role of triplet repeats in disease. The cloning of the disease gene and the identification of expanding repeats represent preliminary steps to the understanding of the full disease process. Genetic testing provides at least an immediate diagnostic tool, but much still remains to be determined for effective therapies to be developed for tomorrow's patients. Many questions, including the biological role of these triplet expansions in evolution, remain unanswered. SEE ALSO ANDROGEN INSENSITIVITY SYNDROME; FRAGILE X SYNDROME; GENE; GENETIC TESTING; INHERITANCE PATTERNS; PLEIOTROPHY.

Nandita Jha

Bibliography

Cummings, Christopher J. "Fourteen and Counting: Unraveling Trinucleotide Repeat Diseases." *Human Molecular Genetics* 9, no. 6 (2000): 909–916.

Green, Howard. "Human Genetic Diseases Due to Codon Reiteration: Relationship to an Evolutionary Mechanism." *Cell* 74 (1993): 955–956.

Moxon, E. Richard, and Christopher Wills. "DNA Microsatellites: Agents of Evolution?" *Scientific American* (January 1999): 94–99.

Perutz, Max F., and A. H. Windle. "Cause of Neural Death in Neurodegenerative Diseases Attributable to Expansion of Glutamine Repeats." *Nature* 412 (2001): 142–144.

Tobin, Allan J., and Ethan R. Signer. "Huntington's Disease: The Challenge for Cell Biologists." *Trends in Cell Biology* 10, no. 12 (2000): 531–536.

Internet Resources

Huntington's Disease Society of America. <www.hdsa.org/>.

International Myotonic Dystrophy Organization. <http://www.myotonicdystrophy.org/>.

Trisomy *See Chromosomal Aberrations*

Tumor Suppressor Genes

mitosis seperation of replicated chromosomes

Tumor suppressor genes regulate **mitosis** and cell division. When their function is impaired, the result is a high rate of uncontrolled cell growth or cancer. Damage to tumor suppressor genes contributes to a large number of different types of **tumors**.

The Balancing Act of Regulating the Cell Cycle

tumors masses of undifferentiated cells; may become cancerous

The **cell cycle** is a fundamental process of life, regulated by a balance of positive and negative mechanisms that act at key points during cell growth and differentiation. Proto-oncogenes tend to "push" the cell cycle (in a positively acting manner) by activating various cell cycle pathways within the developing cell. By contrast, tumor suppressor genes normally repress, or "put the brakes on," the activation of the pathways.

cell cycle sequence of growth, replication, and division that produces new cells

Genetics of Tumor Suppressor Genes

Mutations in tumor suppressor genes can arise spontaneously by exposure to a mutagenic substance such as ultraviolet light or certain chemicals. In such cases, only the mutated cell and its descendants will be affected. Mutations can also be inherited from a parent or arise early in development. In these cases, almost all the cells of the body will inherit the same mutation.

A mutation in a single tumor suppressor gene is usually not enough to cause cancer. This is because each cell contains two copies of each gene, one inherited from each parent. Most cancer-causing mutations cause a loss of function in the mutated gene. Often, having even one functional copy is enough to prevent disease, and two mutations are needed for cancer to develop. This is known as the "two-hit" model of carcinogenesis.

This model was first described in retinoblastoma, a common cancer of the **retina**. The affected gene (called the retinoblastoma gene) is a tumor suppressor. Spontaneous (noninherited) mutations are rare, but since there are many millions of cells in the retina, several of them will develop the gene mutation over the course of a lifetime. It would be very unlikely, though, for a single cell to develop two spontaneous mutations (at least in the absence of prolonged exposure to carcinogens), and thus noninherited retinoblastoma is very rare. When it occurs, it almost always affects only one eye—the eye in which the unlucky doubly hit cell resides.

retina light-sensitive layer at the rear of the eye

If, however, a person inherits one copy of an already mutated gene from one parent, every cell in the eye starts life with one "hit." The chances are very high that several cells will suffer another hit sometime during their life. The chances are thus very high that the person will develop retinoblastoma, almost always in both eyes, since the necessary second hit is common enough that cells in both eyes will be affected. Because inheriting a single copy of the mutated gene is so likely to lead to the disease, the gene is said to show a dominant inheritance pattern.

Generalized Tumor Suppressor Genes

There are a growing number of genes that have been identified as having some function as tumor suppressor genes. The table below lists genes and their associated tumor types. One of the most important tumor suppressor

Table 1.

Gene Symbol	Gene Name	Main Tumor Type	Secondary Tumor Type	Chromosomal Location
APC	Adenomatous polyposis coli	Familial adenomatous polyposis of the colon	—	5q21-q22
BRCA1 and 2	Familial breast/ ovarian cancer 1 and 2	Hereditary breast cancer	—	13q12.3
CDKN1C	Cyclin-dependent kinase inhibitor 1C (p57) gene	Beckwith-Wiedemann syndrome	Wilms' tumor and rhabdomyosarcoma	11p15.5
MEN1	Multiple endocrine	Multiple endocrine neoplasia	Parathyroid/pituitary	11q13
NF1	Neurofibromatosis type 1 gene	Neurofibromatosis type 1 syndrome	Neurofibromas, gliomas, pheochromocytomas and myeloid leukemia	17q11.2
NF2	Neurofibromatosis type 2 gene	Neurofibromatosis type 2 syndrome	Bilateral acoustic neuromas, meningiomas and ependymomas	22q12.2
TSC1	Tuberous sclerosis type 1	Tuberous sclerosis	Some hamartomas and renal cell carcinoma	9q34
TSC2	Tuberous sclerosis type 2	Tuberous sclerosis	Some hamartomas and renal cell carcinoma	16p13.3

Tumor suppressor genes act as gatekeepers for passage from S (synthesis) phase to G2 and mitosis. Activation of these genes, which may occur if DNA replication cannot be successfully completed, triggers apoptosis, or programmed cell death.

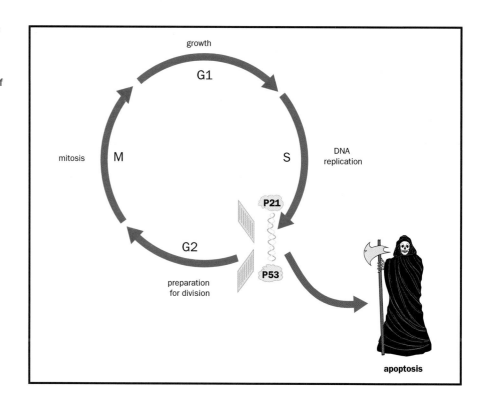

genes is *TP53* (more commonly known as *p53*). This gene was originally identified as a germ-line mutation in the rare inherited cancer called Li-Fraumeni Syndrome, but it has since been shown to be involved in a wide variety of cancer types. The *p53* gene is lost (e.g., the gene is deleted from the chromosome) in about 50 percent of all cancerous cells.

The p53 protein is responsible for controlling the cell cycle checkpoint at the stage where the cell makes a decision to duplicate its genome, called the G2/S boundary. Along with p21 (another essential protein at this boundary), p53 protein monitors the state of the DNA to ensure that the genome is intact and not damaged. The S phase is where the genome is duplicated to get ready for cell division, so it is important that any damage and errors be repaired. If the cell is unable to repair the damage to its DNA, p53 can induce the programmed cell death pathway (called apoptosis) that kills off the cell, thus preventing division of a cell with damaged DNA. If p53 is not functional, the cell cycle is not arrested and any errors will be duplicated and passed on when the cell divides.

Mechanisms of Functional Tumor Suppressor Loss

There are three main ways in which a cell can lose the functionality of its tumor suppressor genes. Chromosomal aberrations, such as balanced reciprocal translocations, can occur. In such translocations, two unlike chromosomes switch segments. The most common such aberration is the chromosome 11 and 22 t(11;22) (q23;q11) translocation. It occurs in 10 to 15 of every 10,000 newborns and is the most common cause of childhood leukemia. The chromosome 9 and 22 t(9;22)(q34;q11) translocation gives rise to the characteristic derivative of chromosome 22, called the Philadelphia chromosome after the city where it was first found, and results in chronic myelogenous leukemia.

Constitutional chromosomal aberrations, which include deletions and **aneuploidy**, are sometimes associated with an increase in specific kinds of cancers. For example, a deletion on chromosome 13 band q14.1 is associated with retinoblastoma. Trisomy 21 in individuals with Down syndrome is associated with a 1 percent occurrence of leukemia.

aneuploidy abnormal chromosome numbers

Viral oncoproteins can interact with tumor suppressor gene proteins. The human papillomavirus (HPV) is a small DNA virus that causes warts. Various subtypes of HPV are associated with cervical cancer. The viral transforming protein E7 has the ability to interact with the retinoblastoma protein, thus interfering with the cell cycle checkpoint controlled by the retinoblastoma protein. Similarly, another HPV gene, *E6*, interacts with the *p53* gene, causing the degradation of the p53 protein, thus allowing the cell cycle to go unchecked. SEE ALSO APOPTOSIS; BREAST CANCER; CANCER; CELL CYCLE; CHROMOSOMAL ABERRATIONS; COLON CANCER; ONCOGENES.

Giles Watts

Bibliography

Rosenberg, S. A., and B. M. John. *The Transformed Cell: Unlocking the Mysteries of Cancer.* New York: Putnam, 1992.

Weinberg, R. A. *Racing to the Beginning of the Road: The Search for the Origin of Cancer.* New York: W. H. Freeman, 1998.

———. *One Renegade Cell: How Cancer Begins.* New York: Basic Books, 1999.

Twins

Twins are siblings carried together in the womb and born at the same time. Similarities and differences between twins can be used to answer questions about the role genes and the environment play in the development of traits such as personality, intelligence, and susceptibility to disease. While results from any single pair of twins cannot provide conclusive answers to such questions, the study of large numbers of twin pairs allows researchers to draw conclusions about inheritance with a significant degree of confidence.

Developmental Mechanisms

Twins are classified as either dizygotic or monozygotic. Dizygotic twins (also called fraternal twins) arise from two separately fertilized eggs, or zygotes. In humans, usually only one egg is released at a time from a woman's ovaries. When two are released, both may become fertilized by separate sperm and implant in the uterus. Dizygotic twins develop separate placentas and amniotic sacs. They may be of the same or different sexes. In the absence of reproductive technology interventions, dizygotic twinning occurs in approximately three of every thousand human births, a rate that increases with maternal age, varies with ethnic group, and is probably influenced by genes that control pituitary function. Various types of assisted reproductive technologies routinely create dizygotic twins, triplets, and higher numbers of offspring.

Monozygotic twins (also called identical twins) arise from a single fertilized egg. At some point after the zygote begins to divide, the cell mass splits into two, creating two embryos from one. Monozygotic twinning

Monozygotic, or identical, twins occur in one-quarter of one percent of all human births and begin life with the same set of genes.

occurs in approximately 0.25 percent of human births. Monozygotic twins are always of the same sex. If the cell mass splits before about day five after fertilization, the two embryos will develop with separate placentas and separate amniotic sacs. This occurs in about two-thirds of human monozygotic twins. Between day five and about day nine, splitting leads to two amniotic sacs but one placenta. This occurs in about one-third of Monozygotic twins. Twins that split after day nine will share the amniotic sac. Splitting that late also increases the likelihood that the twins will not separate completely and will develop into conjoined (Siamese) twins.

Monozygotic versus Dizygotic Twins

Because monozygotic (MZ) twins develop from a single fertilized egg, they begin life with exactly the same set of genes. In this respect, they are clones—organisms whose genes are identical. As discussed below, however, they may accumulate genetic and other differences during development.

In contrast, dizygotic (DZ) twins are no more genetically close than any pair of siblings. While it is commonly said that siblings share half their genes, this is incorrect for two reasons. First, the random nature of meiosis and fertilization means that two siblings could end up with many, or few, genes from a particular parent in common. Second, there are many human genes for which there is only one common form, or allele. Therefore, any two people will share many alleles, regardless of their relationship. Only those genes with more than one allele form the basis of human genetic variation. These are the real focus of the question about gene-sharing in siblings. Of these variable genes, siblings (including dizygotic twins) on average share half.

Because dizygotic twins are the same age, they may share more of their environment than would two siblings of different ages. For instance, because they are likely to be engaged in similar activities, dizygotic twins are more likely to have similar environmental exposures (including behaviors, diet, hobbies, exposure to infectious agents, and exposure to chemicals)—whether at home, at school, or in the community—than two siblings of different ages and different activity patterns. It is this similarity of environment but difference of genes that makes them a useful contrast to monozygotic twins, whose environments and genes are largely identical.

Similarities and Differences between Monozygotic Twins

The fertilized egg cell that gives rise to MZ twins begins life with a single set of genes, and so we might predict that every cell that arises from it would be exactly identical. However, small differences between daughter cells may accumulate throughout embryonic development and later in life. The earliest difference may be in the **mitochondria** each inherits. Mitochondria are the cell's power plants and contain a small amount of DNA. Some of the hundreds of mitochondria in a cell may contain mutations. If the cells that create the two twins carry different mitochondrial genes, even identical twins will be genetically different. Mutations can also accumulate during embryonic development, or after birth, either in the mitochondrial genes or the genes in the nucleus. Such mutations may have a significant effect: Some types of cancer are due to mutations accumulated during one's lifetime, often through exposure to environmental chemicals or radiation.

mitochondria energy-producing cell organelle

For the vast majority of genes, though, MZ twins are exactly identical. Nonetheless, twins do experience slightly different environments, even when reared together, and any early differences between them may be accentuated by families members, or by one another, leading to development of very different personalities.

Amazing Twin Similarities

Some of the most tantalizing clues to the genetic basis of human personality and behavior come from studies of MZ twins reared apart since birth. Such twins have the same genes but, presumably, different environments. A major study of more than 100 such twin pairs showed some remarkable coincidences. A pair of twins meeting for the first time at age thirty-nine each arrived wearing seven rings, two bracelets on one wrist, and a watch and one bracelet on the other wrist. Another twin pair discovered they each had dogs named Toy, had married and divorced women named Linda, remarried women named Betty, and named their sons James Allan and James Alan.

Table 1.

CONCORDANCE IN TWIN STUDIES

Pairwise concordance

$$\frac{\text{Number of twin pairs in which both are affected}}{\text{Total number of twin pairs}}$$

Proband-wise Concordance

$$\frac{[2c_2 + c_1]}{[2c_2 + c_1 + d]}$$

A proband is an independently ascertained twin with the disease; independently ascertained means the twin was NOT identified through the co-twin.

c_2 = the number of concordant pairs in which both twins are probands
c_1 = the number of concordant pairs in which only one twin is a proband
d = the number of discordant pairs

Using concordance patterns to estimate the relative contributions of genetic and environmental determinants to a condition or disorder:

If MZ concordance = 100%

Only genetic determinants likely

If MZ > DZ concordance

Genetic determinants important
Environmental modifiers likely

If MZ concordance = DZ concordance

Shared environmental determinants likely

While these coincidences are amazing, it is important to remember that many more twin pairs in this study did not have such parallel lives or habits. Such stories are curious and provocative but cannot by themselves tell us about the relative contributions of genetics and the environment in shaping personality, behavior, health, or other aspects of the self.

Twin Studies and Concordance

Insight into such questions can be gleaned by several types of studies that compare twins. Comparison of MZ twins reared apart is one type of study but is hampered by the extreme rarity of such twin pairs. Another type of study, comparing MZ twins to DZ twins, is more commonly done, because there are many hundreds of thousands of such twin pairs worldwide. Data on twins have been collected by numerous research groups who have created large and growing databases (registries) that can be mined for information.

Determining a characteristic called concordance plays a crucial role in most such studies. A twin pair is said to be concordant for a trait if both members show it. If neither twin shows the trait, the pair is also concordant, but for the absence of the trait. For instance, twins are concordant for Alzheimer's disease if both develop it. They are discordant if one does have the disease but the other does not.

If a trait is strongly influenced by genes, more MZ twin pairs should be concordant than DZ twin pairs, because MZ twins share more genes. Comparing concordance rates between the two groups, and applying some mathematical analysis, allows researchers to estimate the genetic contribution to a trait, as shown in Table 1.

PAIRWISE CONCORDANCE FOR PARKINSON'S DISEASE

	Concordant Pairs		Discordant Pairs		Pairwise Concordance		Risk of Concordance if MZ
	MZ	DZ	MZ	DZ	MZ	DZ	RR (95% CI)
Overall	11	10	60	80	15.5%	11.1%	1.39 (0.63-3.10)
First twin diagnosed ≤ 50	4	2	0	10	100.0%	16.7%	6.00 (1.69-21.3)
First twin diagnosed > 50	7	8	58	68	10.8%	10.5%	1.02 (0.39-2.67)

Table 2. The data in the last column indicate the relative risk (RR) and 95 percent confidence interval (CI) for concordance of Parkinson's disease in MZ twins. The risk to the second twin is much higher if the first is diagnosed before age fifty, indicating a strong genetic component in early-onset PD.

Twin Registries

Twin studies can have several starting points. Some investigators begin simply by trying to identify twins who will volunteer to be part of a particular research study. Often twins are sought by advertising for twins with the particular disease of interest. This approach has the advantage of simplicity, as twins identify themselves to the research team.

However, twins who volunteer may differ in some important way from those who do not volunteer, and this could affect the conclusions drawn from the study. For example, MZ twins are more likely to volunteer, in general, than DZ twins are. This tendency to volunteer for twin studies among MZ twins is probably because being a twin is a more central part of the identity of MZ pairs than DZ pairs. Also, twins concordant for a particular disease are more likely to volunteer than those without the disease are. If both influences are at work in the same study, more concordant MZ twins than DZ twins may be identified, not because there is an actual difference in concordance between MZ and DZ twins (and thus a genetic effect at work), but because more concordant MZ twins volunteered for the study. If this pattern of volunteerism is mistaken to represent the true pattern of the disease in all twins, an inappropriate conclusion that the disease has genetic causes could result.

Other twin registries attempt to identify all twins within a particular population. One approach is the statewide or national twin registry. All twin births in the region are reported to a central registrar. This results in a more complete picture of all twin pairs in these populations. Examples include the statewide Virginia and Minnesota twin registries in the United States and many national twin registries, including those in the United Kingdom, Australia, the Scandinavian countries, Germany, Belgium, the Netherlands, Italy, and Sri Lanka.

Twin registries have also been assembled from among special populations. Examples in the United States are registries assembled from military records (the World War II Veteran Twins Registry and the Vietnam Era Twin Registry) and from Medicare files (the U.S. Registry of Elderly African-American Twins). In these registries, likely adult twins were identified by searching records to identify individuals with identical dates of birth, birthplaces, and surnames. These individuals were then contacted to verify whether they actually constituted a twin pair. Registries may also be established by identifying twin births within a health maintenance organization (such as the Kaiser Permanente Twin Cohort, in California).

Each registry varies in the amount of contact with registrants. In all, individual contact is strictly monitored to preserve the privacy of each twin. Every research proposal must be approved by a panel to assure the scientific value of the project, the justification for doing the study in twins, and to ensure that the privacy and safety of individual twins will be protected.

Twin registries can be useful starting points for investigating many questions about the genetic and environmental determinants of a trait. Records linkage studies involve no personal contact with the twins. Instead, information in the twin registry is "linked" electronically to information in another database, such as a national health insurance database or a cancer registry. In this way, twins with a particular health problem can be identified, and concordance estimates can be calculated. Similarly, information collected for each twin at registration can later be used to investigate certain kinds of questions without ever contacting the individual twins. On the other end of the spectrum, twins may be asked to volunteer for physical examinations, blood tests, radiological studies, or interviews. Depending on the questions asked, such studies may be useful for comparing concordance, or for identifying risk factors or modifying factors for a trait.

Twin Studies to Investigate the Cause of Parkinson's Disease

An example of the use of investigations in twins to understand more about a disease is provided by recent work in Parkinson's disease. Parkinson's disease (PD) is a progressive neurodegenerative disease causing slowness, tremor, and problems with walking and balance. PD is rare before age fifty but becomes more common thereafter, with increasing age. The cause of PD has long been debated. Both genetic and environmental causes have been suggested, but neither has been definitively shown. Researchers turned to studies in twins to determine the relative contribution of genes and environment to the disease.

The first studies identified twin pairs by recruiting through physicians and PD patient organizations. Studies in the United States, the United Kingdom, and Germany identified 103 pairs, of which only thirteen were concordant for PD. In Finland, forty-two twins with PD were identified by records linkage, but among these was only one concordant pair—a DZ pair. No study had convincingly demonstrated greater monozygotic than dizygotic concordance for the disease, and in all studies the preponderance of twin pairs were discordant for disease. These findings supported an environmental cause of PD. Nonetheless, the advent of molecular genetics prompted great interest in investigations of genetic causes of disease and prompted the resurgence of the hypothesis that all PD had a genetic cause. To address this, a study in a large, unselected cohort—the National Academy of Sciences/National Resource Council (NAS/NRC) World War II Veteran Twins Registry—was undertaken.

In the mid-1950s, the Medical Follow-up Agency of the Institute of Medicine of the NAS/NRC established a registry of approximately 32,000

WHY STUDY TWINS?

- To estimate the relative contributions of genes and environment to the cause of disease by comparing MZ to DZ concordance;
- To investigate environmental determinants of etiology in discordant twin pair studies;
- To investigate environmental influences on disease course in concordant twin pair studies;
- To characterize "presymptomatic" or "at risk" states by studying the unaffected twins in discordant pairs.

Caucasian male twins, all of whom were born between 1917 and 1927 and were veterans of the U.S. Armed Services. In all, 161 twin pairs were identified, twenty-one of which were concordant for PD, as shown in Table 2. In those few pairs with early-onset PD, concordance was greater in MZ pairs. In those with more typical PD, beginning after age fifty, there was no difference in MZ and DZ concordance.

These findings suggest a strong genetic determinant for early-onset disease but predominantly environmental causes in more typical late-onset disease. One caveat is the narrow age range of the twins, who were sixty-seven to seventy-seven years old when studied. Since PD is a late-life disorder, PD in some twins may have been missed with an examination at only one time point. To overcome this, a second evaluation is in progress.

Risk-Factor Investigations in Twins

Studies of twin pairs discordant for disease can be useful for identifying risk factors for disease. Since both genetic and environmental factors are extensively shared by twins, particularly by MZ twins, case-control studies can be particularly powerful. In such a study, each twin is interviewed with regard to specific environmental factors—such as occupation, lifestyle factors, illnesses or injuries, and diet—prior to the onset of the disease in the affected twin. The presence of these factors in the twin with the disease is compared to the twin without disease. An association of an environmental factor with the disease suggests this factor may be causally related. Factors more common among the unaffected twins suggest that the factor may protect against the development of the disease.

Environmental influences on PD have been investigated by studying discordant twin pairs. PD has repeatedly been found to be more common in people who do not smoke cigarettes. Some have proposed that some people are genetically predisposed to both Parkinson's disease and smoking, while others suggest cigarette smoking somehow prevents the degeneration that leads to PD. In two studies of discordant twin pairs, cigarette smoking was more common in the twin without Parkinson's disease, especially in the MZ pairs. Because monozygotic twins are genetically identical, this pattern tips the scales in favor of a direct biological action of cigarette smoke.

As medicine focuses more on early intervention or prevention, it becomes important to identify those persons at risk for a particular condition. This can be a problem if there is no diagnostic test. In discordant twin pairs, the unaffected twin is more likely to be "at risk" for a particular condition, whether due to shared genes or environment, than would be true for two nontwins. Therefore, studying the unaffected "at risk" twin may help to clarify what features are useful for predicting those who later will develop a particular disease. For example, in the PD twin study, the unaffected twins are being studied prospectively with brain imaging tests that may show early evidence of injury to the brain area damaged in PD. If abnormalities on this test are found to precede the development of PD, this could provide a useful method of early detection. When treatments to slow or stop the onset of PD are available, individuals with imaging abnormalities may receive intervention before symptoms develop.

Results from Twin Studies of Other Disorders and Conditions

The twin study method has been used to try to determine the extent of genetic or environmental influence on a wide variety of traits and conditions. Among these are sense of humor, which appears to be largely environmentally determined, as MZ and DZ pairs have similar concordance. Examples of other diseases in which MZ concordance exceeds DZ concordance, suggesting a significant genetic component, include addictive behaviors such as cigarette smoking and alcohol drinking, mental illnesses such as schizophrenia, as well as stroke and certain types of high blood pressure. Twin studies of many other disorders are ongoing.

Conclusion

Twin studies provide a unique approach to investigating the determinants of a disease or condition. A single twin study cannot absolutely determine the importance of genetic or environmental factors. However, the twin study method, in combination with other approaches, can be a powerful tool for unraveling the causes of disease. SEE ALSO BEHAVIOR; FERTILIZATION; GENE AND ENVIRONMENT; GENE DISCOVERY; INHERITANCE PATTERNS.

Caroline M. Tanner and Richard Robinson

Bibliography

Bouchard, T. J., et al. "The Sources of Human Psychological Differences: The Minnesota Study of Twins Reared Apart." *Science* 250 (1990): 223–228.

Segal, Nancy L. *Entwined Lives: Twins and What They Tell Us about Human Behavior.* New York: Plume, 2000.

Wright, Lawrence. *Twins: And What They Tell Us about Who We Are.* New York: John Wiley & Sons, 1997.

Internet Resource

Minnesota Twin Family Study. University of Minnesota. <http://www.psych.umn .edu/psylabs/mtfs/default.htm>.

Viroids and Virusoids

genome the total genetic material in a cell or organism

Viruses are infectious agents consisting of a nucleic acid **genome** made of DNA or RNA, a protein coat, and sometimes lipids. They are able to replicate only inside cells, and the viral genome contains genes coding for proteins. Viroids and virusoids are also infectious agents, but they differ from viruses in several ways. For instance, they have a single-stranded circular, RNA genome. Their genomes are very small and do not code for proteins. Viroids replicate autonomously inside a cell, but virusoids cannot. Rather, virusoid replication requires that the cell is also infected with a virus that supplies "helper" functions.

Viroids

Viroids infect plant cells, and more than twenty-five kinds in two families are known. Viroid RNA is 246 to 375 nucleotides long and it folds to form rodlike structures with nucleotide base pairing (in which A pairs with U, C

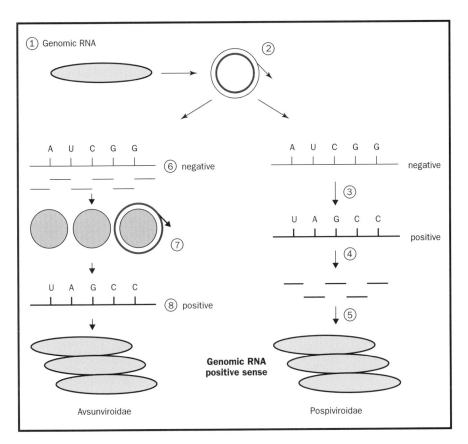

① Genomic RNA

A U C G G
⑥ negative

⑦

U A G C C
⑧ positive

Avsunviroidae

A U C G G
negative

③

U A G C C
positive

④

⑤

Genomic RNA positive sense

Pospiviroidae

②

Pospiviroidae, or virusoid RNA (1), is copied by RNA polymerase (2), gives a negative strand first (3), and then a positive strand (4). This is cleaved and re-circularized to make viroid or virusoid RNA (5). Avsunviroidae, or virusoid RNA, is copied to make negative strand RNA (2, 6). Following ribozyme cleavage the negative circular RNA (7) is copied to make positive strand RNA (8). Ribozyme cleavage and recircularization results in formation of genomic RNA. Negative strand RNA is shown in thin lines, positive strand RNA is shown in thick lines.

pairs with G). The potato spindle tuber viroid and tomato plant macho viroid, both members of the family Pospiviroidae, replicate in the cell nucleus. A cellular enzyme, RNA polymerase, copies the circular RNA of the viroid to make a linear, repeated copy of the genome in **complementary** or "negative sense." This RNA is copied again to make another linear, repeated "positive sense" RNA. Cellular enzymes cut this second copy of RNA at each place where the genome begins a repetition, yielding multiple copies of the genome. These copies then reassume a circular shape to make new viroid RNA.

complementary matching opposite, like hand and glove

The members of the second family of viroids, the Avsunviroidae, replicate in cell chloroplasts. Two are known, the avocado sunblotch viroid and the peach latent mosaic viroid. In both of these species, RNA polymerase makes a long, linear negative sense RNA. This RNA contains a catalytic **ribozyme** sequence, which cleaves itself. The negative sense RNA resumes its customary circular shape and is copied to form linear positive sense RNA. The ribozyme again cleaves this RNA, yielding linear genomic units that again recircularize, forming viroids.

ribozyme RNA-based catalyst

The differing replication strategies of these two groups reflect different evolutionary origins. Viroids move through the plant in the phloem and plasmodesmata, which are part of the plant's circulatory system. They propagate by mechanical means, vegetative reproduction, and possibly via seeds and insects. They cause plant diseases following interactions with proteins. For example, when they interact with an enzyme that impairs protein synthesis, the growth of the host plant may be stunted. This can have severe economic consequences.

Virusoids

Virusoid genomes are 220 to 388 nucleotides long. A virusoid genome does not code for any proteins, but instead serves only to replicate itself. Virusoids can replicate in the **cytoplasm** and possess a ribozyme activity. RNA replication is similar to that of viroids, but each requires that the cell be infected with a specific "helper" virus. Five virusoids are known, and the helper viruses for these are all members of the Sobemovirus family. An example of a "helper" virus is the subterranean clover mottle virus, which has an associated virusoid. Virus enzymes may aid replication of the virusoid RNA. The virusoid is incorporated into the virus particle and transmitted as a "satellite," a separate nucleic acid not part of the viral chromosome. Replication of the helper virus is independent of the virusoid.

Virusoids belong to a larger group of infectious agents called satellite RNAs, which are found in bacteria, plants, fungi, invertebrates, and vertebrates. Some satellite RNAs encode proteins; these are called satellite viruses and, like virusoids, must coinfect with a helper virus in order to replicate. One satellite RNA infecting humans is the hepatitis delta virusoid. It has a circular, single-stranded RNA genome of 1,700 nucleotides. Its helper is the hepatitis B virus, which is associated with liver disease. Coinfection with both agents results in a more severe infection. SEE ALSO RIBOZYME; RNA POLYMERASES; VIRUS.

Shaun Heaphy

cytoplasm the material in a cell, excluding the nucleus

Bibliography

Internet Resource

Viroids and Virusoids. University of Leicester, Department of Microbiology & Immunology. <http://www-micro.msb.le.ac.uk/335/viroids.html>.

Virus

A virus is a parasite that must infect a living cell to reproduce. Although viruses share several features with living organisms, such as the presence of genetic material (DNA or RNA), they are not considered to be alive. Unlike cells, which contain all the structures needed for growth and reproduction, viruses are composed of only an outer coat (capsid), the genome, and, in some cases, a few enzymes. Together these make up the **virion**, or virus particle. Many illnesses in humans, including AIDS, influenza, Ebola fever, the common cold, and certain cancers, are caused by viruses. Viruses also exist that infect animals, plants, bacteria, and fungi.

virion virus particle

Physical Description and Classification

Viruses are distinguished from free-living microbes, such as bacteria and fungi, by their small size and relatively simple structures. Diminutive viruses such as parvovirus may have a diameter of only 25 nanometers (nm, 10^{-9} meters). Poxviruses, the largest known viruses, are about 300 nanometers across, just at the detection limits of the light microscope. Typical bacteria have diameters of 1,000 nanometers or more. Information on the structure of viruses has been obtained with several techniques, including electron

Nucleic Acid	Polarity	Family	Examples	Host	Diseases/pathologies
ss DNA	+	*Parvoviridae*	parvovirus B19	humans	erythema infectiosum (fifth disease)
ds DNA	+/-	*Myoviridae*	Bacteriophage T4	*E. coli*	bacterial lysis
		Papillomaviridae	HPV types 2, 16, 18, 33	humans	warts, cervical and other cancers
		Herpesviridae	herpes zoster virus	humans	chicken pox, shingles
		Poxviridae	variola virus	humans	smallpox
ss RNA non-seg.	+	*Picornaviridae*	poliovirus types 1-3	humans	poliomyelitis
			rhinovirus (100+ serotypes)	humans	common cold
		Togaviridae	equine encephalitis virus	insects/horses	CNS disease in horse and humans
ss RNA non-seg.	-	*Rhabdoviridae*	rabies virus	mammals	rabies
		Paramyxoviridae	measles virus	humans	measles
ssRNA segmented	-	*Orthomyxoviruses*	influenza virus	mammals, birds	influenza
ssRNA segmented	- and/ or ambi	*Bunyaviridae*	Sin Nombre virus	rodents	hanta fever
		Arenaviridae	Lassa fever virus	primates	hemorrhagic fever
ds RNA	+/-	*Reoviridae*	Rice dwarf virus	plants	stunting
ssRNA DNA rep. int.	+	*Retroviridae*	HIV types 1, 2	humans	AIDS
			HTLV type I	humans	adult T-cell leukemia
ds DNA +/- RNA rep. int.	+/-	*Hepadnaviridae*	hepatitis B virus	humans	hepatitis, hepatocellular carcinoma

ss=single-stranded; ds=double stranded; non-seg.= non-segmented; ambi = ambisense; rep. int = replicative intermediate; HPV= human papillomavirus; CNS = central nervous system.

microscopy (EM). The limit of resolution of traditional EM is about 5 nm. With advanced EM techniques, such as cryogenic EM (cryoEM, in which the sample is rapidly frozen instead of exposed to chemical fixatives), coupled with computer image processing, smaller structures (1–2 nm) can be resolved. However, X-ray crystallography is the only method that allows for atomic-level resolution. Small viruses that produce uniform particles can be crystallized. The first atomic-level structure of a virus, tomato bushy stunt virus, was solved in 1978.

Table 1. Classification of selected viruses by nucleic acid replication strategy (Baltimore Scheme).

There is great diversity among viruses, but a limited number of basic designs. Capsids are structures that contain the viral genomes; many have icosahedral symmetry. An icosahedron is a three-dimensional, closed shape composed of twenty equilateral triangles. Viral proteins, in complexes termed "capsomers," form the surface of the icosahedron.

Other viruses, such as the virus that causes rabies, are helical (rod shaped). The length of helical viruses can depend on the length of the genome, the DNA or RNA within, since there are often regular structural interactions between the nucleic acids of the genome and the proteins that cover it.

A **lipid**-containing envelope is a common feature of animal viruses, but uncommon in plant viruses. Embedded in the envelope are surface proteins, usually **glycoproteins** that help the virus interact with the surface of the cell it is infecting. A matrix layer of proteins often forms a bridge between the surface glycoproteins and the capsid. Some viruses, such as the picornaviruses, are not enveloped, nor do they have a matrix layer. In these viruses, cell-surface interactions are mediated by the capsid proteins.

lipid fat or waxlike molecule, insoluble in water

glycoproteins proteins to which sugars are attached

Some viruses have compound structures. The head of the T4 bacterial virus (bacteriophage) is icosahedral and is attached via a collar to a contractile

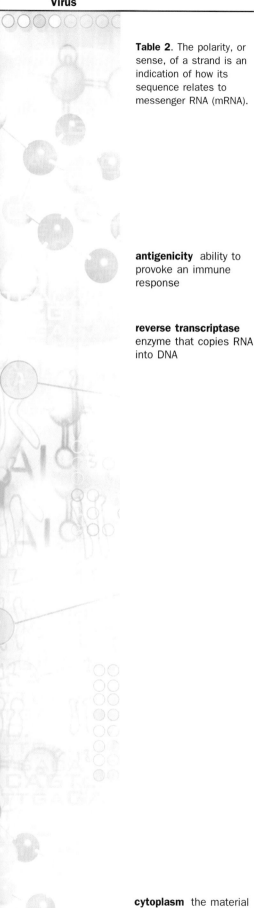

Table 2. The polarity, or sense, of a strand is an indication of how its sequence relates to messenger RNA (mRNA).

Molecule	Sequence	Polarity or Sense
Complementary RNA	A U U G G G C U C	negative
Coding strand DNA	T A A C C C G A G	positive
Complementary DNA	A T T G G G C T C	negative
mRNA	U A A C C C G A G	positive

antigenicity ability to provoke an immune response

reverse transcriptase enzyme that copies RNA into DNA

tail with helical symmetry. Large viruses, such as the herpesviruses and poxviruses, can have higher-ordered and more complex structures.

Classification of viruses considers the genome characteristics, virion shape and macromolecular composition, and other properties, such as **antigenicity** and host range. A scheme for classification of viruses based on the type of nucleic acid (DNA or RNA) present in the virus particle and the method of genome replication was devised by David Baltimore, co-discoverer of **reverse transcriptase** (see Table 1). Reverse transcriptase is an enzyme that converts retroviral genomic single-stranded (ss) RNA into doubled-stranded (ds) DNA.

Viral genomes can be RNA or DNA, positive or negative in polarity, ss or ds, and one continuous (sometimes circular) molecule or divided into segments. By convention, messenger RNA (mRNA) that can be directly translated to protein is considered positive sense (or positive in polarity). DNA with a corresponding sequence (that is, the coding strand of double-stranded DNA) is also a positive-sense strand. An RNA or DNA molecule with the reverse complementary sequence to mRNA is a negative-sense strand. A few viruses have been identified that contain one or more "ambisense" genomic RNA segments that are positive sense in one part of the molecule (this part can be translated directly into protein) and negative sense (reverse complement of coding sequence) in the rest of the molecule.

Virus Replication Cycle

For a virus to multiply it must infect a living cell. All viruses employ a common set of steps in their replication cycle. These steps are: attachment, penetration, uncoating, replication, assembly, maturation, and release.

Attachment and Penetration. A virion surface protein must bind to one or more components of the cell surface, the viral receptors. The presence or absence of receptors generally determines the type of cell in which a virus is able to replicate. This is called viral tropism. For example, the poliovirus receptor is present only on cells of higher primates and then in a limited subset of these, such as intestine and brain cells. While called virus receptors, these are actually used by the cell for its own purposes, but are exploited by the virus for entry.

Entry of the viral genome into the cell can occur by direct penetration of the virion at the cell surface or by a process called endocytosis, which is the engulfment of the particle into a membrane-based vesicle. If the latter, the virus is released when the vesicle is acidified inside the cell. Enveloped viruses may also fuse with the cellular surface membrane, which results in release of the capsid into the **cytoplasm**. Surface proteins of several viruses

cytoplasm the material in a cell, excluding the nucleus

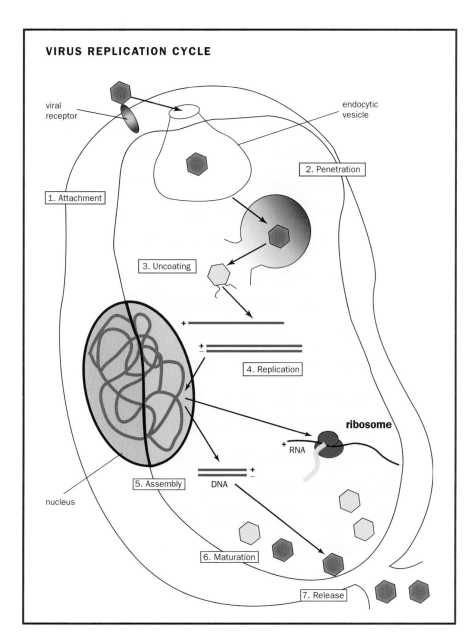

VIRUS REPLICATION CYCLE

viral receptor

endocytic vesicle

1. Attachment

2. Penetration

3. Uncoating

4. Replication

ribosome

+ RNA

5. Assembly

DNA

nucleus

6. Maturation

7. Release

Virus attaches to a receptor, enters the cell, and is uncoated, releasing its genetic material. In this example, positive polarity (+) DNA is released, which triggers formation of a complementary (−) DNA strand. Double-stranded DNA is copied in the nucleus to make progeny DNA. It is also transcribed to make messenger RNA, for synthesis of viral proteins at the ribosome.

contain "fusion peptides," which are capable of interacting with the lipid bilayers of the host cell.

Uncoating and Replication. After penetration, viral capsid proteins must be removed, at least partly, to express and replicate the viral genome. In the case of most DNA viruses, the capsid is routed to the nucleus prior to uncoating. An example can be seen in the poxviruses, whose large DNA genomes encode most of the proteins needed for DNA replication. These viruses uncoat and replicate completely in the cytoplasm. RNA viruses typically lose the protective envelope and capsid proteins upon penetration into the cytoplasm. In reoviruses, only an outer protein shell is removed and replication takes place inside a structured subviral particle.

Viral genomes must be expressed as mRNAs in order to be translated into structural proteins for the capsids and, in some cases, as replicative proteins for replicating the virus genome. Viral genomes must also provide templates

that can be replicated to produce progeny genomes that will be packaged into newly produced virions. Replication details vary among the different types of viruses.

The ss positive-sense DNA of parvoviruses is copied by host DNA polymerase (the enzyme that replicates DNA) in the nucleus into a negative-sense DNA strand. This in turn serves as a template for mRNA and progeny DNA synthesis. The genomes of larger DNA viruses, with the exception of the poxviruses, are also transcribed and replicated in the nucleus by a combination of viral and host enzymes (for example, DNA-dependent RNA polymerses for transcription of mRNAs, DNA-dependent DNA polymerase for genome replication).

Positive-polarity RNA virus genomes can be translated directly, but for effective progeny production additional rounds of RNA replication via a negative-stranded intermediate are required. This is accomplished by a viral transcriptase (RNA-dependent RNA polymerase) and associated cofactors. Single-stranded negative-sense RNA viruses of animals must also carry a viral transcriptase to transcribe functional mRNAs and subsequently produce proteins, since this RNA-to-RNA enzymatic activity is typically lacking in animal cells.

Retroviruses are unique among viruses in that the genome is diploid, meaning that two copies of the positive-polarity RNA genome are in each virus particle. The genomic RNA is not translated into protein, but rather serves as a template for reverse transcription, which produces a double-stranded DNA via a viral reverse transcription enzyme. The DNA is subsequently integrated into the host cell chromosomal DNA. Hepadnaviruses also encode a reverse transcriptase, but replication occurs inside the virus particle producing the particle-associated genomic DNA.

Assembly, Maturation, and Release. As viral proteins and nucleic acids accumulate in the cell, they begin a process of self-assembly. Viral self-assembly was first demonstrated in a seminal series of experiments in 1955, wherein infectious particles of tobacco mosaic virus spontaneously formed when purified coat protein and genomic RNA were mixed. Likewise, poliovirus capsomers are known to self-assemble to form a procapsid in the cytoplasm. Progeny positive-strand poliovirus RNAs then enter this nascent particle. "Chaperone" proteins (chaparonins) of the cell play a critical role in facilitating the assembly of some viruses. Their normal role is to help fold cellular proteins after synthesis.

The maturation and release stages of the replication cycle may occur simultaneously with the previous step, or may follow in either order. Many viruses assemble their various components into "immature" particles. Further intracellular or extracellular processing is required to produce a mature infectious particle. This may involve cleavage of precursors to the structural proteins, as in the case of retroviruses.

Viruses that are not enveloped usually depend upon disintegration or **lysis** of the cell for release. Enveloped viruses can be released from the cell by the process of budding. In this process the viral capsid and usually a matrix layer are directed to a modified patch of cellular membrane. Interactions between the matrix proteins and/or envelope proteins drive envelopment. In the case of viruses that bud at the cell surface, such as some

lysis breakage

togaviruses and retroviruses (including HIV), this also results in release of the virus particles. If the virus acquires a patch of the nuclear membrane (as is the case with herpesviruses), then additional steps involving vesicular transport may be required for the virus to exit the cell.

Infection Outcomes

Viral infection can result in several different outcomes for the virus and the cell. Productive infection, such that each of the seven steps outlined above occurs, results in the formation of progeny viruses. Cells productively infected with poliovirus can yield up to 100,000 progeny virions per cell, although only a small fraction (fewer than 1 per 1,000) of these are capable of going on to carry out a complete replication cycle of their own. Productive infection may induce cell lysis, which results in the death of the cell. Nonenveloped viruses typically induce cell lysis to permit release of progeny virions. Many enveloped viruses also initiate events that result in cell death by various means, including **apoptosis**, **necrosis**, or lysis.

apoptosis programmed cell death

Viral infection may be abortive, in which one or more necessary factors, either viral or cellular, are absent and progeny virions are not made. Infection may be nonproductive, at least transiently, but viral genomes may still become resident in the host cell. Herpesviruses and retroviruses can establish **latent** infections. Latently infected cells may express a limited number of viral products, including those that result in cell transformation. Latent infections can often be activated by various stimuli, such as stress in the case of herpesviruses, to undergo a productive infection.

necrosis cell death from injury or disease

latent present or potential, but not apparent

Viral Cancers

Infection with certain viruses can also result in cell transformation, stable genetic changes in the cell that result in disregulated cell growth and extended growth potential (immortalization). In animals, such virally induced cellular changes can result in cancer. This correlation was first made by Harry Rubin and Howard Temin in the 1950s, when they observed that Rous sarcoma virus, a retrovirus capable of inducing solid tumors in chickens, could also cause biochemical and structural changes and extend the proliferative potential of cultured chicken cells.

Viruses are perhaps second only to tobacco as risk factors for human cancers. DNA tumor viruses include papillomaviruses and various herpes viruses (such as HHV-8, which causes Kaposi's **sarcoma**). More than sixty strains of human papillomaviruses (HPV) have been identified. HPV cause warts, which are benign tumors, but are also the causes of **malignant** penile, vulval, and cervical cancers. Infection with hepatitis B or C viruses is associated with increased incidence of liver cancer. Adenoviruses have been shown to induce cancers in animals, but not in humans. Retroviruses can also cause cancer in various animal species, including humans. HTLV-1 causes adult T-cell leukemia in about 1 percent of infected humans.

sarcoma a type of malignant (cancerous) tumor

malignant cancerous; invasive tumor

Viruses can cause cancer through their effects on two important cellular genes or gene products: tumor suppressors and **oncogenes**. These genes are critical players in cell-cycle regulation. One protein product from HPV binds to the retinoblastoma (Rb) tumor suppressor protein. HPV E6 protein binds p53 tumor suppressor protein and promotes its degradation.

oncogenes genes that cause cancer

RHABDOVIRUS (RABIES VIRUS, VSV)

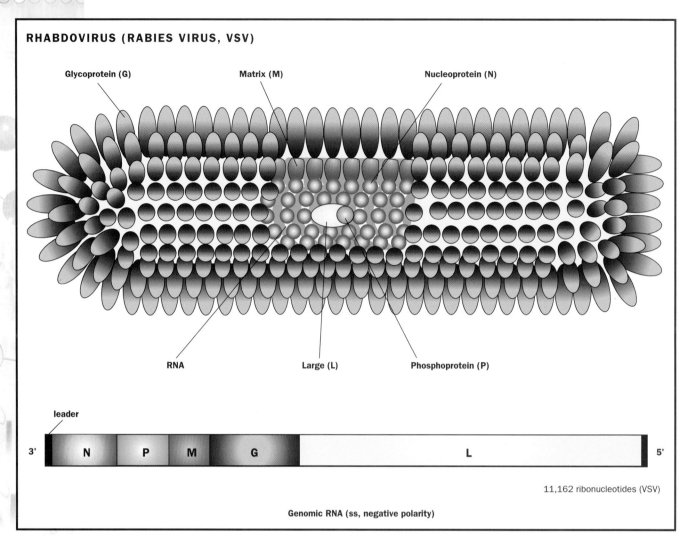

Rhabdoviruses are helical viruses and include the virus that causes rabies and vesicular stomatitis virus (VSV), a common laboratory virus. The lipid-containing envelope is embedded with glycoprotein (G) spikes. A layer of a matrix protein (M) forms a bridge between G and the nucleocapsid proteins (N). Also included in the virion are several molecules of the RNA-dependent RNA polymerase (Large or L protein) and its cofactor, the phosphoprotein (P). The genes for these proteins are arranged as shown in the viral genome.

Acutely transforming retroviruses, which induce tumors in a short time period of weeks to months, carry modified versions of cellular oncogenes, called viral oncogenes. Slowly transforming retroviruses also subvert cellular oncogenes, but by integrating into or near the oncogene, thereby altering its expression, a process that can take years because of the apparently random nature of retrovirus integration.

Vaccines

Many viral infections can be prevented by vaccination. Several classes of vaccines are currently in use in humans and animals. Inactivated vaccines, such as the poliovirus vaccine developed by Jonas Salk, are produced from virulent viruses that are subjected to chemical treatments that result in loss of infectivity without complete loss of antigenicity (antigenicity is the ability to produce immunity). Another approach is the use of weakened variants of a virus with reduced **pathogenicity** to induce a protective immune response

pathogenicity ability to cause disease

without disease. While vaccines are usually given before exposure to a virus, postexposure vaccines can cure some virus infections with extended incubation periods, such as rabies.

Vaccines against smallpox eradicated the illness in 1980. It is believed that it may also be possible to eliminate polio. A recombinant vaccine against hepatitis B virus is now produced in yeast. However, developing effective vaccines to some viruses, including the common cold viruses, HIV-1, herpesviruses, and HPV, is proving very difficult principally due to the existence of many variants. Public health measures, such as mosquito control programs to curb the spread of viral diseases transmitted by these **vectors**, and safe-sex campaigns to slow the spread of sexually transmitted diseases, can also be effective. Because viruses replicate in cells, drugs that target viruses typically also affect cell functions. These therapeutic agents must be active against the virus while having "acceptable toxicity" to the host organism. The majority of the specific antiviral drugs currently in use target viral enzymes. For example, **nucleoside** analogues that target viral polymerases are active against HIV and certain herpesviruses. SEE ALSO BIOTECHNOLOGY; CANCER; CELL, EUKARYOTIC; DNA POLYMERASES; GENE THERAPY; HIV; ONCOGENES; RETROVIRUS; TRANSFORMATION; TUMOR SUPPRESSOR GENES; VIROIDS AND VIRUSOIDS.

vectors carriers

nucleoside building block of DNA or RNA, composed of a base and a sugar

Robert F. Garry

Bibliography

Garrett, Laurie. *The Coming Plague: Newly Emerging Diseases in a World out of Balance.* New York: Penguin Books, 1994.

Kolata, Gina B. *Flu: The Story of the Great Influenza Pandemic of 1918 and the Search for the Virus That Caused It.* New York: Farrar, Straus and Giroux, 1999.

Preston, Richard. *The Hot Zone.* New York: Random House, 1994.

Internet Resource

Sanders, David M., and Robert F. Garry. "All the Virology on the World Wide Web." <www.virology.net>.

Watson, James

Geneticist
1928–

James Dewey Watson was the codiscoverer of the structure of DNA. He has also made major contributions to research in genetics and molecular biology as an administrator, and has written widely read and influential books for both academic and nonscience audiences.

Early Life and Training

Watson was born April 6, 1928, in Chicago, Illinois. He showed his brilliance early, finishing high school in two years and appearing as one of the original "Quiz Kids," on a popular 1940s radio show of the same name. He was graduated from the University of Chicago in 1947 with a B.S. in zoology, reflecting an early love of birds. He did his doctoral work at Indiana University in genetics, and earned a Ph.D. in 1950. He was drawn to Indiana by the chance

to work with Hermann Joseph Muller, who had been one of Thomas Hunt Morgan's associates in the famous "fly room" at Columbia University, and who had received a Nobel Prize for his discoveries in genetics. Watson's thesis adviser and principal mentor was Salvador Luria, who, along with Max Delbrück, had established bacterial genetics as the experimental system in which most of the major discoveries in molecular biology were to be made. Watson's thesis was on the effect of X rays on the multiplication of a bacterial virus, called phage.

Watson continued to study phage as a postdoctoral student in Copenhagen, Denmark where he worked from 1950 to 1951. While there, he met Maurice Wilkins, and for the first time saw the X-ray diffraction images generated in Wilkins's lab by Rosalind Franklin. Watson quickly decided to turn his attention to discovering the structure of important biological molecules, including DNA and proteins. By that time, DNA had been shown to be the genetic molecule, and it was believed that it somehow carried the instructions for making proteins, which actually perform most of the work in a cell.

The Structure of DNA

Luria arranged for Watson to continue his work at the Cavendish Laboratory in Cambridge, England, which was a center for the study of biomolecular structure, and Watson arrived there in late 1951. At the Cavendish, he met Francis Crick, who, after training in physics, had turned his attention to similar structural questions. The two hit it off, and began collaborating on the structure of DNA.

Watson and Crick approached the problem by building models of the four nucleotides known to make up DNA. Each was composed of a sugar called deoxyribose, a phosphate group, and one of four bases, called adenine, thymine, cytosine, and guanine. They knew the sugars and phosphates alternated to form a chain, with the bases projecting off to the side. The X-ray images they had seen suggested the structure was a helix, and offered more information about dimensions as well. They also knew that the biochemist Erwin Chargaff had discovered that the amounts of adenine and thymine in a cell's DNA were equal, as were the amounts of cytosine and guanine.

hydrogen bonding weak bonding between the H of one molecule or group and a nitrogen or oxygen of another

template a master copy

After several failed attempts, more analysis of the X-ray images, and a fortuitous conversation with a biochemist who corrected one of their hypothesized base structures, they developed the correct model. The helix is formed from two opposing strands of sugar phosphates, while the bases project into the center. Weak bonding (called **hydrogen bonding**) between bases holds them together. The key, as Watson and Crick discovered, was that the hydrogen bonds work best when adenine pairs with thymine, and guanine with cytosine, thus explaining Chargaff's ratios. The structure immediately suggested a replication mechanism, in which each side serves as the **template** for the formation of a new copy of the opposing side, and they speculated, correctly, that the sequence of the bases was a code for the sequence of amino acids in proteins. They published their results in 1953, and received the Nobel Prize for physiology and medicine for it 1962, along with Wilkins (Franklin by then had died, and was therefore ineligible for the prize).

Later Accomplishments

Watson remained active in the study of DNA and RNA for a number of years after the publication of the DNA structure. He joined the faculty of Harvard University in 1955, and remained there until 1976. During this time, he wrote an influential textbook, *Molecular Biology of the Gene*, and an enormously popular (and colorful) account of his and Crick's discovery, called *The Double Helix*.

In 1968 Watson became the director of the Cold Spring Harbor Laboratory on Long Island, New York, and he became president of the laboratory in 1994, a position he continues to hold. Watson revitalized this laboratory, helping it become one of the premier genetics research institutions in the world. His organizational drive was also called upon in 1988, when he spearheaded the launch of the U.S. Human Genome Project, dedicated to determining the sequence of the entire three billion bases in the genome. He headed the project from 1988 to 1992.

Throughout his career, Watson has invariably been described as "brash," reflecting his capacity to take on big projects and big ideas, and his enthusiasm for making daring, occasionally outrageous predictions about the causes of an unexplained phenomenon or the direction science will take. Explaining this tendency in relation to his work on DNA, Watson wrote, "A potential key to the secret of life was impossible to push out of my mind. It was certainly better to imagine myself becoming famous than maturing into a stifled academic who had never risked a thought." SEE ALSO CRICK, FRANCIS; DELBRÜCK, MAX; DNA; DNA STRUCTURE AND FUNCTION, HISTORY; MORGAN, THOMAS HUNT; MULLER, HERMANN; NUCLEOTIDE.

Richard Robinson

Bibliography

Judson, Horace F. *The Eight Days of Creation*, expanded edition. Cold Spring Harbor, NY: Cold Spring Harbor Press, 1996.

Watson, James. *The Double Helix: A Personal Account of the Discovery of the Structure of DNA*. New York: New American Library, 1991.

———. *Genes, Girls, and Gamow: After the Double Helix*. New York: Knopf, 2002.

Internet Resource

"Biographical Sketch of James Dewey Watson." <http://nucleus.cshl.org/CSHLlib/archives/jdwbio.htm>.

X Chromosome

The X chromosome occupies an exceptional place in the mammalian **genome**. Together with the Y chromosome, the X chromosome differentiates the sexes. Males have one X chromosome and a Y chromosome and females have two X chromosomes. Because of this fundamental genetic difference, diseases caused by genes located on the X chromosome affect males and females differently and thus present unusual inheritance patterns. Furthermore, equal dosage of expression from genes on the X chromosome is restored between males and females by a special process called X inactivation, in which genes on one of the female X chromosomes are shut down.

genome the total genetic material in a cell or organism

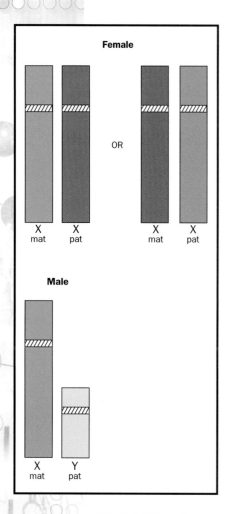

Female

OR

X
mat

X
pat

X
mat

X
pat

Male

X
mat

Y
pat

Figure 1. Schematic representation of the sex chromosomes of a female and male. The active X chromosome is in orange, the inactive X chromosome in purple, and the Y chromsome in green. In females, either the maternal (mat) or paternal (pat) X chromosome is inactivated in any given cell.

recombining exchanging genetic material

germ cells cells creating eggs or sperm

Role of the X Chromosome in Sex Differentiation

The so-called sex chromosomes differentiate the sexes: females are XX and males XY, which is the basis for the development of a fetus into a girl or a boy (Figure 1). All other chromosomes (called autosomes) are present in two copies in both males and females. It is the presence of a Y chromosome that determines the male sex of a baby, because the Y carries a gene that induces undifferentiated gonads to turn into testes in the fetus. The number of X chromosomes does not change the sex of a baby. Indeed, people with a single X chromosome and no Y chromosome are females with Turner syndrome, a rare genetic disorder characterized by short stature and infertility. Conversely, people who have two X chromosomes and a Y chromosome are males with Klinefelter's syndrome, which includes tall stature and infertility.

Sex Chromosome Evolution

Present-day sex chromosomes look very different from each other: The X chromosome comprises about 5 percent of the human genome, and contains about 2,000 genes, while the Y chromosome is quite small and contains only about 50 genes (Figure 1). This striking difference in size and gene content between the sex chromosomes makes it hard to believe that they are actually ancient partners in a pair of chromosomes that originally were very similar. Once sex became determined by a genetic signal from the Y, the sex chromosomes largely stopped **recombining** in **germ cells**. Degeneration of Y genes ensued, together with accumulation of genes that are advantageous to males on the Y chromosome, such as genes involved in testicular function and in male fertility. Similar genes appear to have accumulated on the X chromosome, so that the X chromosome also plays an important role in sperm production. The X chromosome may also have a prominent role in brain function and intelligence. A strong argument in favor of this intriguing but still controversial theory is that mental disability is more common in males.

The X Chromosome and Diseases

Some diseases affect males but not females in a family. Such diseases, called X-linked recessives, are often caused by mutations in genes located on the X chromosome, called X-linked genes. An X-linked disease is transmitted from the mother, not from the father, to an affected male, and an affected male will transmit a copy of the mutant gene to all his daughters. A famous example of an X-linked disease is hemophilia A. The blood of hemophiliac males fails to coagulate properly, leading to thinning of the blood and unstoppable bleeding after injury. This disease was recognized in the royal family of Queen Victoria, where examination of the huge pedigree readily confirmed recessive X-linked inheritance. Only males were affected, having inherited an X chromosome with a copy of a mutated gene from their healthy mothers, who were carriers of the disease. The mutated gene in hemophilia A was identified as factor VIII, a gene that encodes a protein essential for proper clotting of the blood. Males with a mutated gene cannot compensate since they have only one X chromosome, whereas female carriers have one normal gene that can compensate for the diseased gene. This typical recessive X-linked inheritance has been described for a variety of genes.

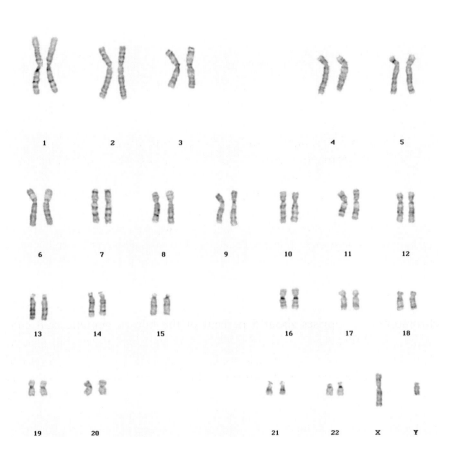

Figure 2. Karyotype of a normal human male. Each of the first twenty-two chromosomes has two copies. The last set is the sex chromosomes, and consists of one X and one Y chromosome.

Dominant X-linked mutations, in which female carriers with just one mutated copy of the gene are affected, are rare. One example of such a disease is vitamin D–resistant rickets, in which people develop skeletal deformities. Generally, the disease is less severe in females than in males, because of X inactivation (see below). A famous X-linked disorder with inheritance that cannot be classified as either recessive or dominant is fragile X mental retardation. The fragile X chromosome bears its name because it displays a site susceptible to chromosome breakage. The mutated gene at the site contains a triplet repeat expansion, in which a series of three consecutive bases are copied multiple times. This causes the gene to be turned off by secondary changes in its structure. Affected males have severe mental retardation and female carriers can also be affected.

X Inactivation

X inactivation consists of the silencing of genes on one of the X chromosomes in the female fetus. This silencing, which results in the absence of protein products from the inactivated genes, restores equal X-linked gene expression between the sexes. So, in the end, females have only one active X chromosome, like males (Figure 1). In the case of individuals with an abnormal number of X chromosomes, such as three X chromosomes, only one X will remain active.

One may wonder then why females do not express **deleterious** recessive X-linked mutations like males. This is because X inactivation is random, and a female is a mosaic of cells with either her paternal X active or her maternal X active (Figure 1). Thanks to this randomness, female carriers usually

deleterious harmful

have plenty of cells with the normal gene remaining functional. Sometimes, there is even cell selection in carrier females, leading to skewed X inactivation in favor of the normal gene remaining functional. One intriguing feature of X inactivation is that it does not affect all X-linked genes. About 20 percent of genes "escape" X inactivation in humans. With a higher expression level in females than in males, such genes could perhaps play a role in female-specific functions. In males, some of these unusual genes have retained a functionally similar gene on the Y, as remnants of the ancient partnership of the sex chromosomes.

X inactivation was discovered in 1961 by Mary Lyon, a British scientist who studied mice. Thus, another name for this phenomenon is "Lyonization." The physiologic or normal regulation of expression of many genes is at the level of the individual gene. In contrast, X inactivation regulates a whole chromosome that comprises a huge number of genes. Special mechanisms of regulation evolved to initiate X inactivation through the action of a master gene on the X. Once one of the two X chromosomes (maternal or paternal) is randomly chosen to become inactivated in a given fetal cell, it will be faithfully maintained in this state in the progeny of the cell. The stability of the inactivation is mediated by a series of complex molecular changes called **epigenetic** modifications. X inactivation is lost in only one type of cells, the female germ cells, where both X chromosomes are functional for transmission to the next generation. Thus, X inactivation involves special mechanisms of initiation, maintenance, and reactivation. Much work still needs to be done to fully understand the fascinating roles of the X chromosome and its regulation. SEE ALSO CHROMOSOMAL ABERRATIONS; FRAGILE X SYNDROME; HEMOPHILIA; INHERITANCE PATTERNS; INTELLIGENCE; MEIOSIS; MOSAICISM; SEX DETERMINATION; Y CHROMOSOME.

Christine M. Disteche

epigenetic not involving DNA sequence change

Bibliography

Miller, Orlando J., and Eeva Therman. *Human Chromosomes.* New York: Springer-Verlag, 2001.

Nussbaum, Robert L., Rod R. McInnes, and Huntington F. Willard. *Thompson & Thompson Genetics in Medicine.* Philadelphia, PA: Saunders, 2001.

Wang, Jeremy P., et al. "An Abundance of X-linked Genes Expressed in Spermatogonia." *Nature Genetics* 27, no. 4 (2001): 422–426.

Y Chromosome

The **diploid** human genome is packaged within 46 chromosomes, as two pairs of 23 discrete elements, into all cells other than the **haploid** gametic egg and sperm cells. During the reproductive process, each parent's gametes contribute 22 nonsex chromosomes and either one X or one Y chromosome.

Paternal Inheritance

The X and Y chromosomes are the sex chromosomes for mammals, including humans. Not only are the X and Y sex chromosomes in mammals physically distinctive, with the Y being smaller, the Y chromosome is exceptionally peculiar. The X chromosome contains considerably more genes than the Y, which has its functionality essentially limited to traits associated with being

diploid possessing pairs of chromosomes, one member of each pair derived from each parent

haploid possessing only one copy of each chromosome

male. It is the Y chromosome that carries the major masculinity-determining gene (*SRY*, for sex-determining region Y), which dictates maleness. In a mating pair, if the paternal partner contributes a normal Y chromosome, male gonadal tissues (testes) develop in the offspring. Only males have the potential to transmit a Y chromosome to the next generation, and thus the father's contribution is decisive regarding an offspring's sex.

Since normally only one Y chromosome exists per cell, no pairing between X and Y occurs at meiosis, except at small regions. Normally, no crossing over occurs. Therefore, except for rare mutations that may occur during spermatogenesis, a son will inherit an identical copy of his father's Y chromosome, and this copy is also essentially identical to the Y chromosomes carried by all his paternal forefathers, across the generations. This is in contrast to the rest of his chromosomal heritage, which will be a unique mosaic of contributions from multiple ancestors created by the reshuffling process of recombination.

Sex Determination and Y Chromosome Genes

While *SRY* is the most dramatic gene affiliated with the Y chromosome, about thirty other genes have been identified. Some notable representatives include *AZFa*, *b*, and *c*, which are associated with spermatogenesis and male infertility, *SMCY*, associated with the immune response function responsible for transplantation rejection when male tissue is grafted to female tissue, and *TSPY*, which may play a role in testicular cancer.

Sex Chromosome Evolution and Peculiarities

Discussions of sex chromosome evolution raise the question of the biological risks and benefits of sexual differentiation in organisms. Overall, sexual **dimorphism** enhances diversity that, in turn, improves the chances for evolutionary change and potential survival during periods of environmental change.

dimorphism two forms

There are risks in the specialization of the Y chromosome, however. Besides its absence in females, lack of recombination for most of its physical territory except at its tips, and the strict pattern of paternal inheritance, the solitary cellular existence of the Y chromosome reduces the opportunity for DNA repair, which normally occurs while pairing during mitosis. This may explain the prevalence of multicopy DNA sequences on the Y, and why many of its genes have lost functionality. In fact, while genes predominately specific to male function tend to accumulate on the Y chromosome, other genes that have functional counterparts elsewhere will atrophy over evolutionary time, through the accumulation of uncorrected mutations. Thus the Y chromosome is slowing evolving toward a composition with fewer and fewer essential genes.

Molecular Anthropology Using the Y Chromosome

The field of molecular anthropology is predicated on the concept that the genes of modern populations encode aspects of human history. By studying the degree of genetic molecular variation in modern organisms, one can, in principle, understand past events. The Y chromosome is uniquely suited to such studies. Secondary applications of Y chromosome variation studies

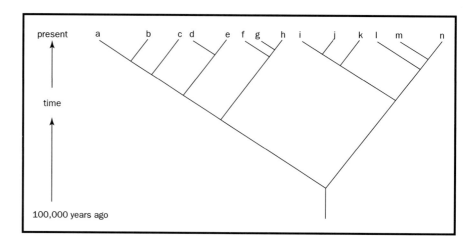

present

time

100,000 years ago

a   b   c d   e f g   h i   j   k l   m   n

Schematic illustration of a gene tree created using Y chromosome polymorphisms. Each modern population (a–n) is descended from a single ancestral populate living 100,000 years ago. Each branching point represents a mutation event, which is then faithfully inherited.

polymorphisms DNA sequence variant

include forensics (criminological investigations, such as determining whether or not an individual has been involved in a crime) and genealogical reconstruction (verifying membership in a particular family's ancestry).

DNA polymers (such as chromosomes) are composed of a four-letter alphabet of chemicals called nucleotide bases. Random unique event mutations in DNA sequences can change the identity of a single base in the DNA molecule. These "spelling changes" are the essential currency of genetic anthropological research.

What is central is the assumption that a particular mutation arose just once in human history, and all men that display such a mutation on their Y chromosome descend from a common forefather on whom the mutation first appeared. The sequential buildup of such mutational events across the generations can be readily determined and displayed as a gene tree. Informally, the last known mutation to accumulate on a particular chromosome can be used to define a particular lineage or branch tip in the tree. As long as the mutational change does not affect the individual's ability to reproduce, it may be preserved and handed down to each succeeding generation, eventually becoming widespread in a population. Such mutations are called **polymorphisms** or genetic markers.

Since most of the Y chromosome has the special property of not recombining during meiosis, no shuffling of DNA from different ancestors occurs. As a consequence, any Y chromosome accumulates all the mutations that have occurred during its lineal life span and thus preserves the paternal genetic legacy that has been transmitted from father to son over the generations. The discovery of numerous Y chromosome polymorphisms has allowed us to deduce a reliable genealogy composed of numerous distinctive lineages. This concept is analogous to the genealogical relationships maintained by the traditional transmission of surnames in some cultures, although the gene tree approach provides access to a prehistorically deeper set of paternal relationships.

Molecular anthropologists have exploited this knowledge in an attempt to understand the history and evolutionary relationships of contemporary populations by performing a systematic survey of Y-chromosome DNA sequence variation. The unique nature of Y-chromosome diversification provides an elegant record of human population histories allowing researchers to reconstruct a global picture, emblematic of modern human origins, affinity, differentiation, and demographic history. The evidence shows that all modern extant human Y chromosomes trace their ancestry to Africa, and that descendants left Africa perhaps less than 100,000 years (or approximately 4,000 generations) ago.

While variation in any single DNA molecule can reflect only a small portion of human diversity, by merging other genetic information, such as data from the maternally transmitted mitochondrial DNA molecule, and nongenetic knowledge derived from archeological, linguistic, and other sources, we can improve our understanding of the affinities and histories of contemporary peoples. SEE ALSO MOLECULAR ANTHROPOLOGY; POLY-MOPHISMS; SEX DETERMINATION; X CHROMOSOME.

Peter A. Underhill

Bibliography

Cavalli-Sforza, Luigi L. *Genes, Peoples, and Languages*, Mark Seielstad, trans. New York: North Point Press, 2000.

Jobling, Mark, and Christopher Tyler-Smith. "New Uses for New Haplotypes: The Human Y Chromosome, Disease, and Selection." *Trends in Genetics* 16 (2000): 356–362.

Strachan, Tom, and Andrew P. Read. *Human Molecular Genetics*. New York: Wiley-Liss, 1996.

Yeast

Yeast are single-celled eukaryotic organisms related to fungi. The baker's yeast *Saccahromyces cerevisiae* and the distantly related *Schizosaccharomyces pombe* are favored model organisms for genetic research. The interest in yeast research stems from the fact that, as eukaryotic organisms, the sub-cellular organization of yeast is similar to that of cells of more complex organisms. Thus, understanding how a particular gene functions in yeast frequently correlates to how similar genes function in mammals, including humans.

Yeast Genetics

Yeast have many advantages as a genetic research tool. First, yeast are non-pathogenic (they do not cause diseases) and are therefore easy and safe to grow. Yeast can divide by simple fission (mitosis) or by budding and, like bacteria, they can be rapidly grown on solid **agar** plates or in liquid **media**. After just a few days in culture, a single yeast cell can produce millions of identical copies of itself, giving scientist a large supply of a genetically pure research tool.

agar gel derived from algae

media nutrient source

Second, yeast grow as either haploids (having only one set of chromosomes) or diploids (with two chromosome sets). Thus, genetically recessive mutations can be readily identified by **phenotypic** (visually observable) changes in the haploid strain. In addition, complementation can be performed by simply mating two haploid strains, where one does not contain the mutation. The resulting diploid strain contains both the functional and nonfunctional version of a gene responsible for a phenotype. The addition

phenotypic related to the observable characteristics of an organism

As shown by a colorized scanning electron micrograph of a yeast cell magnified 3,025 times, this single-celled eukaryotic fungus multiplies by budding.

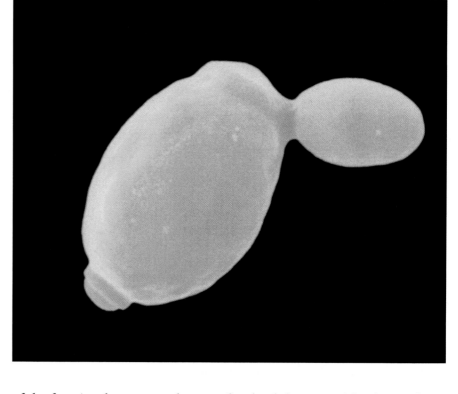

of the functional gene complements for the defect caused by the nonfunctional gene in the haploid strain. Diploid strains can be induced to undergo meiosis, a process in which the cell divides and passes one-half of its chromosomes to each of the resulting cells. After two such divisions, reproductive structures called asci are produced that contain four haploid offspring, called ascospores. The asci can be dissected and each of the ascospores isolated. In this way, scientists can easily mate different yeast strains and obtain new haploid genotypes through sexual reproduction and meiosis.

Third, the genome of yeast is small, about 3.5 times larger than that of bacteria and 200 times smaller than that of mammals. The yeast genome is arranged in 16 linear chromosomes that range from 200 to 2,200 kilobases in length. Unlike mammals, the yeast genome is very compact, with only 12 million base pairs, very few **introns**, and very little spacer DNA between functional genes. As a result, in 1996 baker's yeast was the first eukaryotic organism to have its entire genome sequenced.

introns untranslated portions of genes that interrupt coding regions

Genetic Transformation

Finally, one of the most useful properties of yeast for genetic studies is the ease with which DNA can be introduced into them, in a process called transformation. The introduced DNA can be maintained on self-replicating, circular strands of DNA called plasmids, or it can integrate into the yeast genome. Most importantly, integration usually occurs by a process called homologous recombination, whereby the introduced DNA replaces chromosomal DNA that contains the same sequence. This process permits scientists to readily mutate any yeast gene and replace the native gene in the cells with the mutated version. Since yeast can be grown as haploids, the phenotypic changes caused by the introduced gene can be readily identified.

In addition, the function of a cloned piece of DNA (e.g., a gene) can be identified by transforming yeast in which the DNA is carried on a circular plasmid. The introduced gene may either functionally replace a defective gene or cause a phenotypic defect in the cells indicating a function for that gene.

The ability to complement yeast defects with cloned pieces of DNA has been extended to mammalian genes. Recognizing that some genes have similar sequences and functions in both mammals and yeast, scientists sometimes use yeast as a tool to identify the functions of mammalian genes. Not many mammalian genes can directly substitute for a yeast gene, however. More frequently, scientists study the yeast gene itself to understand how its protein functions in the cell. The knowledge gained can often lead to an understanding of how similar genes might function in mammals. Now that the yeast genome has been completely sequenced and the results have been deposited in a public databank for all to use, rapid progress is being made in identifying all yeast genes and their functions.

An important method for studying mammalian genes in yeast is called the two-hybrid system. This system is used to determine if two proteins functionally interact with each other. Both genes are cloned into yeast plasmids and transformed into the cells. A special detection system is used that is active only when both cloned proteins physically contact each other in the cell. When that happens, scientist can identify which proteins need to interact with each other in order to function.

Yeast are also being used in the laboratory and commercial production of important nonyeast proteins. Foreign genes are transformed into yeast and, after **transcription** and translation, the foreign proteins can be isolated. Because of the ease of growing large quantities of cells, yeast can produce a large amount of the protein. While similar protein production can be performed by bacteria, eukaryotic proteins often do not function when made in bacteria. This is because most eukaryotic proteins are normally altered after translation by the addition of short sugar chains, and these modifications are often required for proper function, but bacteria do not carry out these necessary post-translational modifications. Yeast, however, does permit these modifications, and is thus more likely to produce a functional protein. SEE ALSO CELL, EUKARYOTIC; CELL CYCLE; GENOME; HUMAN GENOME PROJECT; MODEL ORGANISMS; PLASMID; POST-TRANSLATIONAL CONTROL; TRANSFORMATION; TRANSGENIC ANIMALS.

transcription messenger RNA formation from a DNA sequence

Suzanne Bradshaw

Bibliography

Sherman, Fred. "Getting Started with Yeast." In *Methods in Enzymology*, vol. 194, Christine Guthrie and Gerald R. Fink, eds. New York: Academic Press, 1991.

Watson, James D., Michael Gilman, Jan Witowski, and Mark Zoller. *Recombinant DNA*. New York: Scientific American Books, 1992.

Zebrafish

The zebrafish (*Brachydanio rerio*) is a small tropical freshwater fish that began to be used as a genetic model system in the early 1980s. The zebrafish shares numerous anatomical and genetic similarities with higher vertebrates,

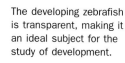

The developing zebrafish is transparent, making it an ideal subject for the study of development.

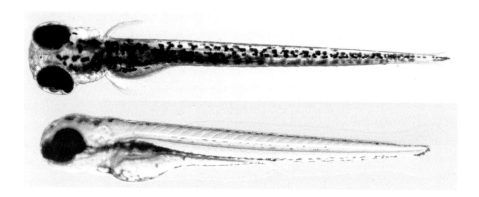

embryogenesis development of the embryo from a fertilized egg

histological related to tissues

including humans, both in the general body plan and in specific organs. Close parallels exist in many aspects of early **embryogenesis** and in the anatomical and **histological** features of the brain, spinal cord, sensory systems, cardiovascular system, and other organs. Not infrequently, genetic defects in zebrafish resemble human disorders. Owing to these similarities, zebrafish genetics is being broadly applied to address both basic biological questions and to model human inherited diseases.

Useful Traits for Researchers

Several characteristics favor the choice of zebrafish for genetic research. First, zebrafish are easy to maintain in large numbers in a small laboratory space. Second, their generation time is relatively short: three months. Finally, females produce large clutches of offspring, about 50 to 100 a week. The zebrafish also presents advantages for embryological analysis: Its embryos develop externally, and are largely transparent for the first 36 hours of development.

The first proponents of the zebrafish model, George Streisinger and his colleagues at the University of Oregon, outlined its genetic characteristics and provided a thorough description of zebrafish embryogenesis. Zebrafish research entered a new phase when two groups, at the University of Tuebingen (Germany), and at Harvard Medical School, performed large-scale chemical **mutagenesis** experiments and isolated nearly 2,000 mutations that affect almost every aspect of embryonic development, from **gastrulation** to axonal pathfinding. Once the mutations were identified, they were studied to determine their developmental effects, after which their genetic and molecular nature was analyzed. These early genetic studies in zebrafish were limited by the lack of genomic resources, such as maps of the genome or genomic libraries, but these deficiencies were gradually eliminated during the late 1990s. Zebrafish research has entered a new phase since the completion of the genome project. This effort provided the sequence of the entire zebrafish genome and made cloning of zebrafish genes much more efficient.

mutagenesis creation of mutations

gastrulation embryonic stage at which primitive gut is formed

Ongoing Research Involving Zebrafish

Zebrafish research continues at a fast pace. New rounds of mutation identification are in progress. To supplement chemical mutagenesis, retroviral vectors were developed as mutagenic agents and applied on a large scale. Tools to study the functions of individual genes, such as **transgenics**, were

transgenics transfer of genes from one organism into another

developed in parallel to mutant screening approaches. A technique that complements mutagenesis screens in zebrafish very well is known as gene knockdown. This approach uses modified antisense oligonucleotides to block the function of specific genes. These oligonuculeotides contain a substitution of the sugar ring in the nucleic acid backbone that makes them resistant to degradation by enzymes in living tissues. Gene knockdown is widely used to study mutant **phenotypes** of genes for which chemically induced mutations are not available.

phenotypes observable characteristics of an organism

The usefulness of the zebrafish as a model organism originates in its unique combination of genetic and embryological characteristics. Genetic approaches, such as mutagenesis screens, can be combined in zebrafish with other techniques, enabling researchers to study cell movements, cell birth dates, or interactions between cells in the living embryo. Although most of the zebrafish genetic research focuses on embryonic development, other problems, such as the genetic basis of circadian rhythms, cancer formation, neurodegenerative disorders, and drug addiction, are also being addressed. SEE ALSO DNA LIBRARIES; MODEL ORGANISMS; MUTAGENESIS; TRANSGENIC ORGANISMS: ETHICAL ISSUES.

Jarema Malicki

Bibliography

Amsterdam, A., and N. Hopkins. "Retrovirus Mediated Insertional Mutagenesis in Zebrafish." *Methods of Cell Biology* 60 (1999): 87–98.

Driever, W., et al. "A Genetic Screen for Mutations Affecting Embryogenesis in Zebrafish." *Development* 123 (1996): 37–46.

Haffter, P., et al. "The Identification of Genes with Unique and Essential Functions in the Development of the Zebrafish, *Danio rerio*." *Development* 123 (1996): 1–36.

Kimmel, C. B., et al. "Stages of Embryonic Development of the Zebrafish." *Development Dynamics* 203 (1995): 253–310.

Malicki, Jarema. "Harnessing the Power of Forward Genetics: Analysis of Neuronal Diversity and Patterning in the Zebrafish Retina." *Trends in Neuroscience* 23 (2000): 531–541.

Nasevicius, A., and S. C. Ekker. "Effective Targeted Gene 'Knockdown' in Zebrafish." *Nature Genetics* 26 (2000): 457.

Thisse, C., and L. Zon. "Organogenesis-Heart and Blood Formation from the Zebrafish Point of View." *Science* 295 (2002): 216–220.

Westerfield, M. *The Zebrafish Book*. Eugene: University of Oregon Press, 1994.

Internet Resource

Zebrafish Information Network. <http://zfin.org>.

Photo Credits

Unless noted below or within its caption, the illustrations and tables featured in *Genetics* were developed by Richard Robinson, and rendered by GGS Information Services. The photographs appearing in the text were reproduced by permission of the following sources:

Volume 1

Accelerated Aging: Progeria (p. 2), Photo courtesy of The Progeria Research Foundation, Inc. and the Barnett Family; *Aging and Life Span* (p. 8), Fisher, Leonard Everett, Mr.; *Agricultural Biotechnology* (p. 10), © Keren Su/Corbis; *Alzheimer's Disease* (p. 15), AP/Wide World Photos; *Antibiotic Resistance* (p. 27), © Hank Morgan/Science Photo Library, Photo Researchers, Inc.; *Apoptosis* (p. 32), © Microworks/Phototake; *Arabidopsis thaliana* (p. 34), © Steinmark/ Custom Medical Stock Photo; *Archaea* (p. 38), © Eurelios/Phototake; *Behavior* (p. 47), © Norbert Schafer/Corbis; *Bioinformatics* (p. 53), © T. Bannor/Custom Medical Stock Photo; *Bioremediation* (p. 60), Merjenburgh/ Greenpeace; *Bioremediation* (p. 61), AP/Wide World Photos; *Biotechnology and Genetic Engineering, History of* (p. 71), © Gianni Dagl Orti/Corbis; *Biotechnology: Ethical Issues* (p. 67), © AFP/Corbis; *Birth Defects* (p. 78), AP/Wide World Photos; *Birth Defects* (p. 80), © Siebert/ Custom Medical Stock Photo; *Blotting* (p. 88), © Custom Medical Stock Photo; *Breast Cancer* (p. 90), © Custom Medical Stock Photo; *Carcinogens* (p. 98), © Custom Medical Stock Photo; *Cardiovascular Disease* (p. 102), © B&B Photos/Custom Medical Stock Photo; *Cell, Eukaryotic* (p. 111), © Dennis Kunkel/ Phototake; *Chromosomal Aberrations* (p. 122), © Pergement, Ph.D./Custom Medical Stock

Photo; *Chromosomal Banding* (p. 126), Courtesy of the Cytogenetics Laboratory, Indiana University School of Medicine; *Chromosomal Banding* (p. 127), Courtesy of the Cytogenetics Laboratory, Indiana University School of Medicine; *Chromosomal Banding* (p.128), Courtesy of the Cytogenetics Laboratory, Indiana University School of Medicine; *Chromosome, Eukaryotic* (p. 137), Photo Researchers, Inc.; *Chromosome, Eukaryotic* (p. 136), © Becker/Custom Medical Stock Photo; *Chromosome, Eukaryotic* (p. 133), Courtesy of Dr. Jeffrey Nickerson/University of Massachusetts Medical School; *Chromosome, Prokaryotic* (p. 141), © Mike Fisher/Custom Medical Stock Photo; *Chromosomes, Artificial* (p. 145), Courtesy of Dr. Huntington F. Williard/University Hospitals of Cleveland; *Cloning Organisms* (p. 163), © Dr.Yorgos Nikas/Phototake; *Cloning: Ethical Issues* (p. 159), AP/Wide World Photos; *College Professor* (p. 166), © Bob Krist/Corbis; *Colon Cancer* (p. 169), © Albert Tousson/Phototake; *Colon Cancer* (p. 167), © G-I Associates/Custom Medical Stock Photo; *Conjugation* (p. 183), © Dennis Kunkel/Phototake; *Conservation Geneticist* (p. 191), © Annie Griffiths Belt/ Corbis; *Delbrück, Max* (p. 204), Library of Congress; *Development, Genetic Control of* (p. 208), © JL Carson/Custom Medical Stock Photo; *DNA Microarrays* (p. 226), Courtesy of James Lund and Stuart Kim, Standford University; *DNA Profiling* (p. 234), AP/Wide World Photos; *DNA Vaccines* (p. 254), Penny Tweedie/Corbis-Bettmann; *Down Syndrome* (p. 257), © Custom Medical Stock Photo.

Volume 2

Embryonic Stem Cells (p. 4), Courtesy of Dr. Douglas Strathdee/University of Edinburgh, Department of Neuroscience; *Embryonic Stem Cells* (p. 5), Courtesy of Dr. Douglas Strathdee/University of Edinburgh, Department of Neuroscience; *Escherichia coli* (*E. coli* bacterium) (p. 10), © Custom Medical Stock Photo; *Eubacteria* (p. 14), © Scimat/Photo Researchers; *Eubacteria* (p. 12), © Dennis Kunkel/Phototake; *Eugenics* (p. 19), American Philosophical Society; *Evolution, Molecular* (p. 22), OAR/National Undersea Research Program (NURP)/National Oceanic and Atmospheric Administration; *Fertilization* (p. 34), © David M. Phillips/Photo Researchers, Inc.; *Founder Effect* (p. 37), © Michael S. Yamashita/Corbis; *Fragile X Syndrome* (p. 41), © Siebert/Custom Medical Stock Photo; *Fruit Fly:* Drosophila (p. 44), © David M. Phillips, Science Source/Photo Researchers, Inc.; *Gel Electrophoresis* (p. 46), © Custom Medical Stock Photo; *Gene Therapy* (p. 75), AP/Wide World; *Genetic Counseling* (p. 88), © Amethyst/Custom Medical Stock Photo; *Genetic Testing* (p. 98), © Department of Clinical Cytogenetics, Addenbrookes Hospital/Science Photo Library/Photo Researchers, Inc.; *Genetically Modified Foods* (p. 109), AP/Wide World; *Genome* (p. 113), Raphael Gaillarde/Getty Images; *Genomic Medicine* (p. 119), © AFP/Corbis; *Growth Disorders* (p. 131), Courtesy Dr. Richard Pauli/U. of Wisconsin, Madison, Clinical Genetics Center; *Hemoglobinopathies* (p. 137), © Roseman/Custom Medical Stock Photo; *Heterozygote Advantage* (p. 147), © Tania Midgley/Corbis; *HPLC: High-Performance Liquid Chromatography* (p. 167), © T. Bannor/Custom Medical Stock Photo; *Human Genome Project* (p. 175), © AFP/Corbis; *Human Genome Project* (p. 176), AP/Wide World; *Individual Genetic Variation* (p. 192), © A. Wilson/Custom Medical Stock Photo; *Individual Genetic Variation* (p. 191), © A. Lowrey/Custom Medical Stock Photo; *Inheritance, Extranuclear* (p. 196), © ISM/Phototake; *Inheritance Patterns* (p. 206), photograph by Norman Lightfoot/National Audubon Society Collection/Photo Researchers, Inc.; *Intelligence* (p. 208), AP/Wide World Photos.

Volume 3

Laboratory Technician (p. 2), Mark Tade/Getty Images; *Maize* (p. 9), Courtesy of Agricultural Research Service/USDA; *Marker Systems* (p. 16), Custom Medical Stock Photo; *Mass Spectrometry* (p. 19), Ian Hodgson/© Rueters New Media; *McClintock, Barbara* (p. 21), AP/Wide World Photos; *McKusick, Victor* (p. 23), The Alan Mason Chesney Medical Archives of The Johns Hopkins Medical Institutions; *Mendel, Gregor* (p. 30), Archive Photos, Inc.; *Metabolic Disease* (p. 38), AP/Wide World Photos; *Mitochondrial Diseases* (p. 53), Courtesy of Dr. Richard Haas/University of California, San Diego, Department of Neurosciences; *Mitosis* (p. 58), J. L. Carson/Custom Medical Stock Photo; *Model Organisms* (p. 61), © Frank Lane Picture Agency/Corbis; *Molecular Anthropology* (p. 66), © John Reader, Science Photo Library/PhotoResearchers, Inc.; *Molecular Biologist* (p. 71), AP/Wide World Photos; *Morgan, Thomas Hunt* (p. 73), © Bettmann/Corbis; *Mosaicism* (p. 78), Courtesy of Carolyn Brown/Department of Medical Genetics of University of British Columbia; *Muller, Hermann* (p. 80), Library of Congress; *Muscular Dystrophy* (p. 85), © Siebert/Custom Medical Stock Photo; *Muscular Dystrophy* (p. 84), © Custom Medical Stock Photo; *Nature of the Gene, History* (p. 103), Library of Congress; *Nature of the Gene, History* (p. 102), Archive Photos, Inc.; *Nomenclature* (p. 108), Courtesy of Center for Human Genetics/Duke University Medical Center; *Nondisjunction* (p. 110), © Gale Group; *Nucleotide* (p. 115), © Lagowski/Custom Medical Stock Photo; *Nucleus* (p. 120), © John T. Hansen, Ph.D./Phototake; *Oncogenes* (p. 129), Courtesy of National Cancer Institute; *Pharmacogenetics and Pharmacogenomics* (p. 145), AP/Wide World Photos; *Plant Genetic Engineer* (p. 150), © Lowell Georgia/Corbis; *Pleiotropy* (p. 154), © Custom Medical Stock Photo; *Polyploidy* (p. 165), AP/Wide World Photos; *Population Genetics* (p. 173), AP/Wide World Photos; *Population Genetics* (p. 172), © JLM Visuals; *Prenatal Diagnosis* (p. 184), © Richard T. Nowitz/Corbis; *Prenatal Diagnosis* (p. 186), © Brigham Narins; *Prenatal Diagnosis* (p. 183), © Dr. Yorgos Nikas/Phototake; *Prion* (p. 188), AP/Wide World Photos; *Prion* (p. 189), AP/

Wide World Photos; *Privacy* (p. 191), © K. Beebe/Custom Medical Stock Photo; *Proteins* (p. 199), NASA/Marshall Space Flight Center; *Proteomics* (p. 207), © Geoff Tompkinson/ Science Photo Library, National Aubodon Society Collection/Photo Researchers, Inc.; *Proteomics* (p. 206), Lagowski/Custom Medical Stock Photo.

Volume 4

Quantitative Traits (p. 2), Photo by Peter Morenus/Courtesy of University of Connecticut; *Reproductive Technology* (p. 20), © R. Rawlins/Custom Medical Stock Photo; *Reproductive Technology: Ethical Issues* (p. 27), AP/Wide World Photos; *Restriction Enzymes* (p. 32), © Gabridge/Custom Medical Stock Photo; *Ribosome* (p. 43), © Dennis Kunkel/ Phototake; *Rodent Models* (p. 61), © Reuters NewMedia Inc./Corbis; *Roundworm:*

Caenorhabditis elegans (p. 63), © J. L. Carson/Custom Medical Stock Photo; *Sanger, Fred* (p. 65), © Bettmann/Corbis; *Sequencing DNA* (p. 74), © T. Bannor/Custom Medical Stock Photo; *Severe Combined Immune Deficiency* (p. 76), © Bettmann/Corbis; *Tay-Sachs Disease* (p. 100), © Dr. Charles J. Ball/ Corbis; *Transgenic Animals* (p. 125), Courtesy Cindy McKinney, Ph.D./Penn State University's Transgenic Mouse Facility; *Transgenic Organisms: Ethical Issues* (p. 131), © Daymon Hartley/Greenpeace; *Transgenic Plants* (p. 133), © Eurelios/Phototake; *Transplantation* (p. 140), © Reuters New Media/Corbis; *Twins* (p. 156), © Dennis Degnan/Corbis; *X Chromosome* (p. 175), © Gale Group; *Yeast* (p. 180), © Dennis Kunkel; *Zebrafish* (p. 182), Courtesy of Dr. Jordan Shin, Cardiovascular Research Center, Massachusetts General Hospital.

Glossary

α the Greek letter alpha

β the Greek letter beta

γ the Greek letter gamma

λ the Greek letter lambda

σ the Greek letter sigma

E. coli the bacterium *Escherichia coli*

"-ase" suffix indicating an enzyme

acidic having the properties of an acid; the opposite of basic

acrosomal cap tip of sperm cell that contains digestive enzymes for penetrating the egg

adenoma a tumor (cell mass) of gland cells

aerobic with oxygen, or requiring it

agar gel derived from algae

agglutinate clump together

aggregate stick together

algorithm procedure or set of steps

allele a particular form of a gene

allelic variation presence of different gene forms (alleles) in a population

allergen substance that triggers an allergic reaction

allolactose "other lactose"; a modified form of lactose

amino acid a building block of protein

amino termini the ends of a protein chain with a free NH_2 group

amniocentesis removal of fluid from the amniotic sac surrounding a fetus, for diagnosis

amplify produce many copies of, multiply

anabolic steroids hormones used to build muscle mass

anaerobic without oxygen or not requiring oxygen

androgen testosterone or other masculinizing hormone

anemia lack of oxygen-carrying capacity in the blood

aneuploidy abnormal chromosome numbers

angiogenesis growth of new blood vessels

anion negatively charged ion

anneal join together

anode positive pole

anterior front

antibody immune-system protein that binds to foreign molecules

antidiuretic a substance that prevents water loss

antigen a foreign substance that provokes an immune response

antigenicity ability to provoke an immune response

apoptosis programmed cell death

Archaea one of three domains of life, a type of cell without a nucleus

archaeans members of one of three domains of life, have types of cells without a nucleus

aspirated removed with a needle and syringe

aspiration inhalation of fluid or solids into the lungs

association analysis estimation of the relationship between alleles or genotypes and disease

asymptomatic without symptoms

ATP adenosine triphosphate, a high-energy compound used to power cell processes

ATPase an enzyme that breaks down ATP, releasing energy

attenuation weaken or dilute

atypical irregular

autoimmune reaction of the immune system to the body's own tissues

autoimmunity immune reaction to the body's own tissues

autosomal describes a chromosome other than the X and Y sex-determining chromosomes

autosome a chromosome that is not sex-determining (not X or Y)

axon the long extension of a nerve cell down which information flows

bacteriophage virus that infects bacteria

basal lowest level

base pair two nucleotides (either DNA or RNA) linked by weak bonds

basic having the properties of a base; opposite of acidic

benign type of tumor that does not invade surrounding tissue

binding protein protein that binds to another molecule, usually either DNA or protein

biodiversity degree of variety of life

bioinformatics use of information technology to analyze biological data

biolistic firing a microscopic pellet into a biological sample (from biological/ballistic)

biopolymers biological molecules formed from similar smaller molecules, such as DNA or protein

biopsy removal of tissue sample for diagnosis

biotechnology production of useful products

bipolar disorder psychiatric disease characterized by alternating mania and depression

blastocyst early stage of embryonic development

brackish a mix of salt water and fresh water

breeding analysis analysis of the offspring ratios in breeding experiments

buffers substances that counteract rapid or wide pH changes in a solution

Cajal Ramon y Cajal, Spanish neuroanatomist

carcinogens substances that cause cancer

carrier a person with one copy of a gene for a recessive trait, who therefore does not express the trait

catalyst substance that speeds a reaction without being consumed (e.g., enzyme)

catalytic describes a substance that speeds a reaction without being consumed

catalyze aid in the reaction of

cathode negative pole

cDNA complementary DNA

cell cycle sequence of growth, replication and division that produces new cells

centenarian person who lives to age 100

centromere the region of the chromosome linking chromatids

cerebrovascular related to the blood vessels in the brain

cerebrovascular disease stroke, aneurysm, or other circulatory disorder affecting the brain

charge density ratio of net charge on the protein to its molecular mass

chemotaxis movement of a cell stimulated by a chemical attractant or repellent

chemotherapeutic use of chemicals to kill cancer cells

chloroplast the photosynthetic organelle of plants and algae

chondrocyte a cell that forms cartilage

chromatid a replicated chromosome before separation from its copy

chromatin complex of DNA, histones, and other proteins, making up chromosomes

ciliated protozoa single-celled organism possessing cilia, short hair-like extensions of the cell membrane

circadian relating to day or day length

cleavage hydrolysis

cleave split

clinical trials tests performed on human subjects

codon a sequence of three mRNA nucleotides coding for one amino acid

Cold War prolonged U.S.-Soviet rivalry following World War II

colectomy colon removal

colon crypts part of the large intestine

complementary matching opposite, like hand and glove

conformation three-dimensional shape

congenital from birth

conjugation a type of DNA exchange between bacteria

cryo-electron microscope electron microscope that integrates multiple images to form a three-dimensional model of the sample

cryopreservation use of very cold temperatures to preserve a sample

cultivars plant varieties resulting from selective breeding

cytochemist chemist specializing in cellular chemistry

cytochemistry cellular chemistry

cytogenetics study of chromosome structure and behavior

cytologist a scientist who studies cells

cytokine immune system signaling molecule

cytokinesis division of the cell's cytoplasm

cytology the study of cells

cytoplasm the material in a cell, excluding the nucleus

cytosol fluid portion of a cell, not including the organelles

de novo entirely new

deleterious harmful

dementia neurological illness characterized by impaired thought or awareness

demography aspects of population structure, including size, age distribution, growth, and other factors

denature destroy the structure of

deoxynucleotide building block of DNA

dimerize linkage of two subunits

dimorphism two forms

diploid possessing pairs of chromosomes, one member of each pair derived from each parent

disaccharide two sugar molecules linked together

dizygotic fraternal or nonidentical

DNA deoxyribonucleic acid

domains regions

dominant controlling the phenotype when one allele is present

dopamine brain signaling chemical

dosage compensation equalizing of expression level of X-chromosome genes between males and females, by silencing one X chromosome in females or amplifying expression in males

ecosystem an ecological community and its environment

ectopic expression expression of a gene in the wrong cells or tissues

electrical gradient chemiosmotic gradient

electrophoresis technique for separation of molecules based on size and charge

eluting exiting

embryogenesis development of the embryo from a fertilized egg

endangered in danger of extinction throughout all or a significant portion of a species' range

endogenous derived from inside the organism

endometriosis disorder of the endometrium, the lining of the uterus

endometrium uterine lining

endonuclease enzyme that cuts DNA or RNA within the chain

endoplasmic reticulum network of membranes within the cell

endoscope tool used to see within the body

endoscopic describes procedure wherein a tool is used to see within the body

endosymbiosis symbiosis in which one partner lives within the other

enzyme a protein that controls a reaction in a cell

epidemiologic the spread of diseases in a population

epidemiologists people who study the incidence and spread of diseases in a population

epidemiology study of incidence and spread of diseases in a population

epididymis tube above the testes for storage and maturation of sperm

epigenetic not involving DNA sequence change

epistasis suppression of a characteristic of one gene by the action of another gene

epithelial cells one of four tissue types found in the body, characterized by thin sheets and usually serving a protective or secretory function

Escherichia coli common bacterium of the human gut, used in research as a model organism

estrogen female horomone

et al. "and others"

ethicists a person who writes and speaks about ethical issues

etiology causation of disease, or the study of causation

eubacteria one of three domains of life, comprising most groups previously classified as bacteria

eugenics movement to "improve" the gene pool by selective breeding

eukaryote organism with cells possessing a nucleus

eukaryotic describing an organism that has cells containing nuclei

ex vivo outside a living organism

excise remove; cut out

excision removal

exogenous from outside

exon coding region of genes

exonuclease enzyme that cuts DNA or RNA at the end of a strand

expression analysis whole-cell analysis of gene expression (use of a gene to create its RNA or protein product)

fallopian tubes tubes through which eggs pass to the uterus

fermentation biochemical process of sugar breakdown without oxygen

fibroblast undifferentiated cell normally giving rise to connective tissue cells

fluorophore fluorescent molecule

forensic related to legal proceedings

founder population

fractionated purified by separation based on chemical or physical properties

fraternal twins dizygotic twins who share 50 percent of their genetic material

frontal lobe one part of the forward section of the brain, responsible for planning, abstraction, and aspects of personality

gamete reproductive cell, such as sperm or egg

gastrulation embryonic stage at which primitive gut is formed

gel electrophoresis technique for separation of molecules based on size and charge

gene expression use of a gene to create the corresponding protein

genetic code the relationship between RNA nucleotide triplets and the amino acids they cause to be added to a growing protein chain

genetic drift evolutionary mechanism, involving random change in gene frequencies

genetic predisposition increased risk of developing diseases

genome the total genetic material in a cell or organism

genomics the study of gene sequences

genotype set of genes present

geothermal related to heat sources within Earth

germ cell cell creating eggs or sperm

germ-line cells giving rise to eggs or sperm

gigabase one billion bases (of DNA)

glucose sugar

glycolipid molecule composed of sugar and fatty acid

glycolysis the breakdown of the six-carbon carbohydrates glucose and fructose

glycoprotein protein to which sugars are attached

Golgi network system in the cell for modifying, sorting, and delivering proteins

gonads testes or ovaries

gradient a difference in concentration between two regions

Gram negative bacteria bacteria that do not take up Gram stain, due to membrane structure

Gram positive able to take up Gram stain, used to classify bacteria

gynecomastia excessive breast development in males

haploid possessing only one copy of each chromosome

haplotype set of alleles or markers on a short chromosome segment

hematopoiesis formation of the blood

hematopoietic blood-forming

heme iron-containing nitrogenous compound found in hemoglobin

hemolysis breakdown of the blood cells

hemolytic anemia blood disorder characterized by destruction of red blood cells

hemophiliacs a person with hemophilia, a disorder of blood clotting

herbivore plant eater

heritability proportion of variability due to genes; ability to be inherited

heritability estimates how much of what is observed is due to genetic factors

heritable genetic

heterochromatin condensed portion of chromosomes

heterozygote an individual whose genetic information contains two different forms (alleles) of a particular gene

heterozygous characterized by possession of two different forms (alleles) of a particular gene

high-throughput rapid, with the capacity to analyze many samples in a short time

histological related to tissues

histology study of tissues

histone protein around which DNA winds in the chromosome

homeostasis maintenance of steady state within a living organism

homologous carrying similar genes

homologues chromosomes with corresponding genes that pair and exchange segments in meiosis

homozygote an individual whose genetic information contains two identical copies of a particular gene

homozygous containing two identical copies of a particular gene

hormones molecules released by one cell to influence another

hybrid combination of two different types

hybridization (molecular) base-pairing among DNAs or RNAs of different origins

hybridize to combine two different species

hydrogen bond weak bond between the H of one molecule or group and a nitrogen or oxygen of another

hydrolysis splitting with water

hydrophilic "water-loving"

hydrophobic "water hating," such as oils

hydrophobic interaction attraction between portions of a molecule (especially a protein) based on mutual repulsion of water

hydroxyl group chemical group consisting of -OH

hyperplastic cell cell that is growing at an increased rate compared to normal cells, but is not yet cancerous

hypogonadism underdeveloped testes or ovaries

hypothalamus brain region that coordinates hormone and nervous systems

hypothesis testable statement

identical twins monozygotic twins who share 100 percent of their genetic material

immunogenicity likelihood of triggering an immune system defense

immunosuppression suppression of immune system function

immunosuppressive describes an agent able to suppress immune system function

in vitro "in glass"; in lab apparatus, rather than within a living organism

in vivo "in life"; in a living organism, rather than in a laboratory apparatus

incubating heating to optimal temperature for growth

informed consent knowledge of risks involved

insecticide substance that kills insects

interphase the time period between cell divisions

intra-strand within a strand

intravenous into a vein

intron untranslated portion of a gene that interrupts coding regions

karyotype the set of chromosomes in a cell, or a standard picture of the chromosomes

kilobases units of measure of the length of a nucleicacid chain; one kilobase is equal to 1,000 base pairs

kilodalton a unit of molecular weight, equal to the weight of 1000 hydrogen atoms

kinase an enzyme that adds a phosphate group to another molecule, usually a protein

knocking out deleting of a gene or obstructing gene expression

laparoscope surgical instrument that is inserted through a very small incision, usually guided by some type of imaging technique

latent present or potential, but not apparent

lesion damage

ligand a molecule that binds to a receptor or other molecule

ligase enzyme that repairs breaks in DNA

ligate join together

linkage analysis examination of co-inheritance of disease and DNA markers, used to locate disease genes

lipid fat or wax-like molecule, insoluble in water

loci/locus site(s) on a chromosome

longitudinally lengthwise

lumen the space within the tubes of the endoplasmic reticulum

lymphocytes white blood cells

lyse break apart

lysis breakage

macromolecular describes a large molecule, one composed of many similar parts

macromolecule large molecule such as a protein, a carbohydrate, or a nucleic acid

macrophage immune system cell that consumes foreign material and cellular debris

malignancy cancerous tissue

malignant cancerous; invasive tumor

media (bacteria) nutrient source

meiosis cell division that forms eggs or sperm

melanocytes pigmented cells

meta-analysis analysis of combined results from multiple clinical trials

metabolism chemical reactions within a cell

metabolite molecule involved in a metabolic pathway

metaphase stage in mitosis at which chromosomes are aligned along the cell equator

metastasis breaking away of cancerous cells from the initial tumor

metastatic cancerous cells broken away from the initial tumor

methylate add a methyl group to

methylated a methyl group, CH_3, added

methylation addition of a methyl group, CH_3

microcephaly reduced head size

microliters one thousandth of a milliliter

micrometer 1/1000 meter

microsatellites small repetitive DNA elements dispersed throughout the genome

microtubule protein strands within the cell, part of the cytoskeleton

miscegenation racial mixing

mitochondria energy-producing cell organelle

mitogen a substance that stimulates mitosis

mitosis separation of replicated chromosomes

molecular hybridization base-pairing among DNAs or RNAs of different origins

molecular systematics the analysis of DNA and other molecules to determine evolutionary relationships

monoclonal antibodies immune system proteins derived from a single B cell

monomer "single part"; monomers are joined to form a polymer

monosomy gamete that is missing a chromosome

monozygotic genetically identical

morphologically related to shape and form

morphology related to shape and form

mRNA messenger RNA

mucoid having the properties of mucous

mucosa outer covering designed to secrete mucus, often found lining cavities and internal surfaces

mucous membranes nasal passages, gut lining, and other moist surfaces lining the body

multimer composed of many similar parts

multinucleate having many nuclei within a single cell membrane

mutagen any substance or agent capable of causing a change in the structure of DNA

mutagenesis creation of mutations

mutation change in DNA sequence

nanometer 10^{-9}(exp) meters; one billionth of a meter

nascent early-stage

necrosis cell death from injury or disease

nematode worm of the Nematoda phylum, many of which are parasitic

neonatal newborn

neoplasms new growths

neuroimaging techniques for making images of the brain

neurological related to brain function or disease

neuron nerve cell

neurotransmitter molecule released by one neuron to stimulate or inhibit a neuron or other cell

non-polar without charge separation; not soluble in water

normal distribution distribution of data that graphs as a bell-shaped curve

Northern blot a technique for separating RNA molecules by electrophoresis and then identifying a target fragment with a DNA probe

Northern blotting separating RNA molecules by electrophoresis and then identifying a target fragment with a DNA probe

nuclear DNA DNA contained in the cell nucleus on one of the 46 human chromosomes; distinct from DNA in the mitochondria

nuclear membrane membrane surrounding the nucleus

nuclease enzyme that cuts DNA or RNA

nucleic acid DNA or RNA

nucleoid region of the bacterial cell in which DNA is located

nucleolus portion of the nucleus in which ribosomes are made

nucleoplasm material in the nucleus

nucleoside building block of DNA or RNA, composed of a base and a sugar

nucleoside triphosphate building block of DNA or RNA, composed of a base and a sugar linked to three phosphates

nucleosome chromosome structural unit, consisting of DNA wrapped around histone proteins

nucleotide a building block of RNA or DNA

ocular related to the eye

oncogene gene that causes cancer

oncogenesis the formation of cancerous tumors

oocyte egg cell

open reading frame DNA sequence that can be translated into mRNA; from start sequence to stop sequence

opiate opium, morphine, and related compounds

organelle membrane-bound cell compartment

organic composed of carbon, or derived from living organisms; also, a type of agriculture stressing soil fertility and avoidance of synthetic pesticides and fertilizers

osmotic related to differences in concentrations of dissolved substances across a permeable membrane

ossification bone formation

osteoarthritis a degenerative disease causing inflammation of the joints

osteoporosis thinning of the bone structure

outcrossing fertilizing between two different plants

oviduct a tube that carries the eggs

ovulation release of eggs from the ovaries

ovules eggs

ovum egg

oxidation chemical process involving reaction with oxygen, or loss of electrons

oxidized reacted with oxygen

pandemic disease spread throughout an entire population

parasites organisms that live in, with, or on another organism

pathogen disease-causing organism

pathogenesis pathway leading to disease

pathogenic disease-causing

pathogenicity ability to cause disease

pathological altered or changed by disease

pathology disease process

pathophysiology disease process

patient advocate a person who safeguards patient rights or advances patient interests

PCR polymerase chain reaction, used to amplify DNA

pedigrees sets of related individuals, or the graphic representation of their relationships

peptide amino acid chain

peptide bond bond between two amino acids

percutaneous through the skin

phagocytic cell-eating

phenotype observable characteristics of an organism

phenotypic related to the observable characteristics of an organism

pheromone molecule released by one organism to influence another organism's behavior

phosphate group PO_4 group, whose presence or absence often regulates protein action

phosphodiester bond the link between two nucleotides in DNA or RNA

phosphorylating addition of phosphate group (PO_4)

phosphorylation addition of the phosphate group PO_4^{3-}

phylogenetic related to the evolutionary development of a species

phylogeneticists scientists who study the evolutionary development of a species

phylogeny the evolutionary development of a species

plasma membrane outer membrane of the cell

plasmid a small ring of DNA found in many bacteria

plastid plant cell organelle, including the chloroplast

pleiotropy genetic phenomenon in which alteration of one gene leads to many phenotypic effects

point mutation gain, loss, or change of one to several nucleotides in DNA

polar partially charged, and usually soluble in water

pollen male plant sexual organ

polymer molecule composed of many similar parts

polymerase enzyme complex that synthesizes DNA or RNA from individual nucleotides

polymerization linking together of similar parts to form a polymer

polymerize to link together similar parts to form a polymer

polymers molecules composed of many similar parts

polymorphic occurring in several forms

polymorphism DNA sequence variant

polypeptide chain of amino acids

polyploidy presence of multiple copies of the normal chromosome set

population studies collection and analysis of data from large numbers of people in a population, possibly including related individuals

positional cloning the use of polymorphic genetic markers ever closer to the unknown gene to track its inheritance in CF families

posterior rear

prebiotic before the origin of life

precursor a substance from which another is made

prevalence frequency of a disease or condition in a population

primary sequence the sequence of amino acids in a protein; also called primary structure

primate the animal order including humans, apes, and monkeys

primer short nucleotide sequence that helps begin DNA replication

primordial soup hypothesized prebiotic environment rich in life's building blocks

probe molecule used to locate another molecule

procarcinogen substance that can be converted into a carcinogen, or cancer-causing substance

procreation reproduction

progeny offspring

prokaryote a single-celled organism without a nucleus

promoter DNA sequence to which RNA polymerase binds to begin transcription

promutagen substance that, when altered, can cause mutations

pronuclei egg and sperm nuclei before they fuse during fertilization

proprietary exclusively owned; private

proteomic derived from the study of the full range of proteins expressed by a living cell

proteomics the study of the full range of proteins expressed by a living cell

protists single-celled organisms with cell nuclei

protocol laboratory procedure

protonated possessing excess H^+ ions; acidic

pyrophosphate free phosphate group in solution

quiescent non-dividing

radiation high energy particles or waves capable of damaging DNA, including X rays and gamma rays

recessive requiring the presence of two alleles to control the phenotype

recombinant DNA DNA formed by combining segments of DNA, usually from different types of organisms

recombining exchanging genetic material

replication duplication of DNA

restriction enzyme an enzyme that cuts DNA at a particular sequence

retina light-sensitive layer at the rear of the eye

retroviruses RNA-containing viruses whose genomes are copied into DNA by the enzyme reverse transcriptase

reverse transcriptase enzyme that copies RNA into DNA

ribonuclease enzyme that cuts RNA

ribosome protein-RNA complex at which protein synthesis occurs

ribozyme RNA-based catalyst

RNA ribonucleic acid

RNA polymerase enzyme complex that creates RNA from DNA template

RNA triplets sets of three nucleotides

salinity of, or relating to, salt

sarcoma a type of malignant (cancerous) tumor

scanning electron microscope microscope that produces images with depth by bouncing electrons off the surface of the sample

sclerae the "whites" of the eye

scrapie prion disease of sheep and goats

segregation analysis statistical test to determine pattern of inheritance for a trait

senescence a state in a cell in which it will not divide again, even in the presence of growth factors

senile plaques disease

serum (pl. sera) fluid portion of the blood

sexual orientation attraction to one sex or the other

somatic nonreproductive; not an egg or sperm

Southern blot a technique for separating DNA fragments by electrophoresis and then identifying a target fragment with a DNA probe

Southern blotting separating DNA fragments by electrophoresis and then identifying a target fragment with a DNA probe

speciation the creation of new species

spindle football-shaped structure that separates chromosomes in mitosis

spindle fiber protein chains that separate chromosomes during mitosis

spliceosome RNA-protein complex that removes introns from RNA transcripts

spontaneous non-inherited

sporadic caused by new mutations

stem cell cell capable of differentiating into multiple other cell types

stigma female plant sexual organ

stop codon RNA triplet that halts protein synthesis

striatum part of the midbrain

subcutaneous under the skin

sugar glucose

supercoiling coiling of the helix

symbiont organism that has a close relationship (symbiosis) with another

symbiosis a close relationship between two species in which at least one benefits

symbiotic describes a close relationship between two species in which at least one benefits

synthesis creation

taxon/taxa level(s) of classification, such as kingdom or phylum

taxonomical derived from the science that identifies and classifies plants and animals

taxonomist a scientist who identifies and classifies organisms

telomere chromosome tip

template a master copy

tenets generally accepted beliefs

terabyte a trillion bytes of data

teratogenic causing birth defects

teratogens substances that cause birth defects

thermodynamics process of energy transfers during reactions, or the study of these processes

threatened likely to become an endangered species

topological describes spatial relations, or the study of these relations

topology spatial relations, or the study of these relations

toxicological related to poisons and their effects

transcript RNA copy of a gene

transcription messenger RNA formation from a DNA sequence

transcription factor protein that increases the rate of transcription of a gene

transduction conversion of a signal of one type into another type

transgene gene introduced into an organism

transgenics transfer of genes from one organism into another

translation synthesis of protein using mRNA code

translocation movement of chromosome segment from one chromosome to another

transposable genetic element DNA sequence that can be copied and moved in the genome

transposon genetic element that moves within the genome

trilaminar three-layer

triploid possessing three sets of chromosomes

trisomics mutants with one extra chromosome

trisomy presence of three, instead of two, copies of a particular chromosome

tumor mass of undifferentiated cells; may become cancerous

tumor suppressor genes cell growths

tumors masses of undifferentiated cells; may become cancerous

vaccine protective antibodies

vacuole cell structure used for storage or related functions

van der Waal's forces weak attraction between two different molecules

vector carrier

vesicle membrane-bound sac

virion virus particle

wet lab laboratory devoted to experiments using solutions, cell cultures, and other "wet" substances

wild-type most common form of a trait in a population

Wilm's tumor a cancerous cell mass of the kidney

X ray crystallography use of X rays to determine the structure of a molecule

xenobiotic foreign biological molecule, especially a harmful one

zygote fertilized egg

Topic Outline

Genetically Modified Foods
HPLC: High-Performance Liquid Chromatography
Pharmaceutical Scientist
Plant Genetic Engineer
Polymerase Chain Reaction
Recombinant DNA
Restriction Enzymes
Reverse Transcriptase
Transgenic Animals
Transgenic Microorganisms
Transgenic Organisms: Ethical Issues
Transgenic Plants

CAREERS

Attorney
Bioinformatics
Clinical Geneticist
College Professor
Computational Biologist
Conservation Geneticist
Educator
Epidemiologist
Genetic Counselor
Geneticist
Genomics Industry
Information Systems Manager
Laboratory Technician
Microbiologist
Molecular Biologist
Pharmaceutical Scientist
Physician Scientist
Plant Genetic Engineer
Science Writer
Statistical Geneticist
Technical Writer

CELL CYCLE

Apoptosis
Balanced Polymorphism
Cell Cycle
Cell, Eukaryotic
Centromere
Chromosome, Eukaryotic
Chromosome, Prokaryotic
Crossing Over
DNA Polymerases
DNA Repair
Embryonic Stem Cells
Eubacteria
Inheritance, Extranuclear

Linkage and Recombination
Meiosis
Mitosis
Oncogenes
Operon
Polyploidy
Replication
Signal Transduction
Telomere
Tumor Suppressor Genes

CLONED OR TRANSGENIC ORGANISMS

Agricultural Biotechnology
Biopesticides
Biotechnology
Biotechnology: Ethical Issues
Cloning Organisms
Cloning: Ethical Issues
Gene Targeting
Model Organisms
Patenting Genes
Reproductive Technology
Reproductive Technology: Ethical Issues
Rodent Models
Transgenic Animals
Transgenic Microorganisms
Transgenic Organisms: Ethical Issues
Transgenic Plants

DEVELOPMENT, LIFE CYCLE, AND NORMAL HUMAN VARIATION

Aging and Life Span
Behavior
Blood Type
Color Vision
Development, Genetic Control of
Eye Color
Fertilization
Genotype and Phenotype
Hormonal Regulation
Immune System Genetics
Individual Genetic Variation
Intelligence
Mosaicism
Sex Determination
Sexual Orientation
Twins
X Chromosome
Y Chromosome

Archaea
Cell, Eukaryotic
Eubacteria
Evolution, Molecular
Fruit Fly: *Drosophila*
HIV
Maize
Model Organisms
Nucleus
Prion
Retrovirus
Rodent Models
Roundworm: *Caenorhabditis elegans*
Signal Transduction
Viroids and Virusoids
Virus
Yeast
Zebrafish

POPULATION GENETICS AND EVOLUTION

Antibiotic Resistance
Balanced Polymorphism
Conservation Biologist
Conservation Biology: Genetic Approaches
Evolution of Genes
Evolution, Molecular
Founder Effect
Gene Flow
Genetic Drift
Hardy-Weinberg Equilibrium

Heterozygote Advantage
Inbreeding
Individual Genetic Variation
Molecular Anthropology
Population Bottleneck
Population Genetics
Population Screening
Selection
Speciation

RNA

Antisense Nucleotides
Blotting
DNA Libraries
Genetic Code
HIV
Nucleases
Nucleotide
Reading Frame
Retrovirus
Reverse Transcriptase
Ribosome
Ribozyme
RNA
RNA Interference
RNA Polymerases
RNA Processing
Transcription
Translation

Cumulative Index

Page numbers in **boldface type** indicate article titles; those in *italic type* indicate illustrations or caption text. Terms are often defined multiple times; page numbers are given for only the first definition in each volume. The number preceding the colon indicates volume number.

α, defined, 3:212

α-helices
    conversion to β-sheets, 3:188–190, 3:*189*, 3:*190*
    protein structure role, 3:200–201, 3:*203*

α-protobacteria, as mitochondrial ancestor, 2:195, 3:57

β, defined, 1:46, 1:216, 3:212

β-globin genes, 1:46

β-sheets
    conversion of α-helices, 3:188–190, 3:*189*, 3:*190*
    protein structure role, 3:200–201, 3:*203*

ΔF508 mutation, 1:*200*, 1:202

2-D PAGE (two-dimensional polyacrylamide gel electrophoresis), 3:207–208

5-HT$_{1b}$ mice, 1:5

A

A3243G gene, diabetes and, 3:54

ABI (Applied Biosystems, Inc.), automated sequencers, 1:44–45

ABL1 gene (Abelson murine strain leukemia viral homolog), 1:95

ABMG (American Board of Medical Genetics), clinical geneticist certification, 1:149, 1:150–151

ABO blood group system
    Bombay phenotype, 2:8, 2:9
    genotypes, 2:128
    inheritance patterns, 2:200, 2:207
    multiple alleles, 1:83–84, 3:82

Aborigines (Australian), eugenic fitness ranking, 2:17

Acceptor (A) site, in translation, 4:138

Accreditation Council for Graduate Medical Education (ACGME), clinical geneticist certification, 1:150–151

Accutane (isotretinoin), 1:80–81

ACE (angiotensin-converting enzyme), 1:8

Acetaldehyde, and alcoholism, 1:5

Acetylation
    of DNA, gene repression role, 3:77
    of proteins, post-translational, 3:203, 3:205, 4:111

Achondroplasia, 1:75–76, 2:130, 2:*131*, 2:202

Achromat color vision defects, 1:172

Acidic activation proteins, 2:64, 2:*66*

Acidity
    Archaea adaptations to, 1:36–37
    defined, 3:200
    impact on protein structure, 3:201

Acids, as chelators, 1:60

Acquired immune deficiency syndrome (AIDS). *See* HIV/AIDS

Acrometageria, progeroid aspects, 1:2

Acrosome cap
    defined, 4:21
    function, 2:35

Acrylamide gels, 1:86, 2:47

ACTH, pituitary gland, 2:160

Actin proteins
    defined, 2:29
    pseudogenes for, 2:30, 3:213

Actinomyces, genome characteristics, 2:116

Activation sites. *See* Enhancer DNA sequences

Activator proteins, 4:111

AD. *See* Alzheimer's disease

ADA gene, SCID role, 4:75

ADA (Americans with Disabilities Act), genetic discrimination and, 2:93

Adaptation, defined, 4:67

Adaptive radiation, 4:92

Addiction, 1:**4–6**
    and birth defects, 1:77
    defined, 1:4
    drug dependence, 1:4, 1:77, 1:81, 3:214
    fruit fly models, 1:5–6
    inheritance patterns, 1:4–5
    rodent models, 1:5
    twin studies, 4:162
    See also Alcoholism; Smoking (tobacco)

Additive effects of genes, 1:178–179, 1:*178*

Adenine
    depurination, 1:240
    in DNA alphabet, 2:83, 4:106
    and DNA structure, 1:215–220, 1:250–251, 2:50, 2:*51*, 3:94
    evolution of, 2:22–24
    mutagenic base analogs, 3:87
    origin of term, 1:249
    polyadenylated repeats, 2:53, 4:9, 4:145
    reactions with aflatoxins, 1:245
    in RNA alphabet, 4:46–47, 4:106
    structure, 1:*216*, 3:115, 3:*116*, 3:*118*, 3:119, 3:*119*, 4:47, 4:*48*, 4:*49*
    See also Base pairs

Adeno-associated virus, as gene therapy vector, 2:76

Adenoma, defined, 1:167

Adenoma-carcinoma sequence, 1:167

Adenomatous cells, cancerous tumors, 1:94

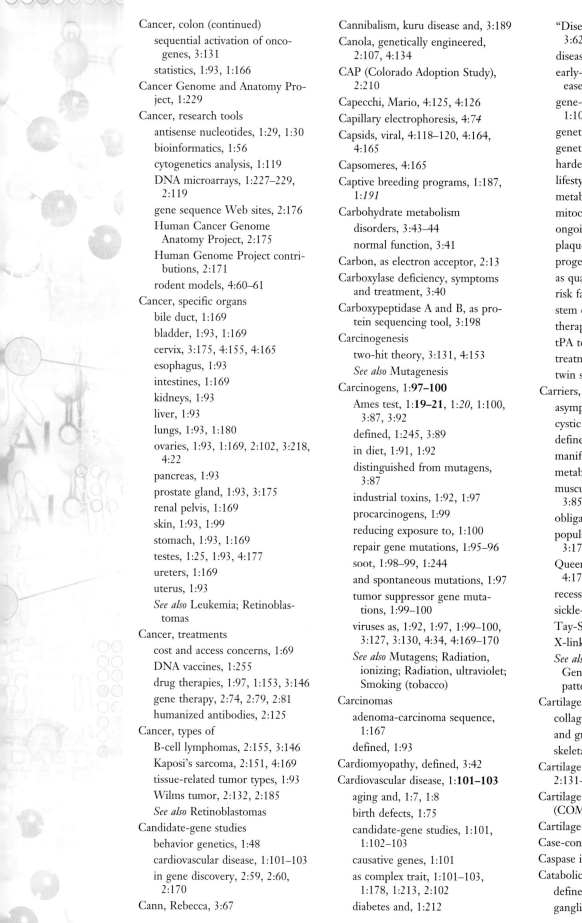

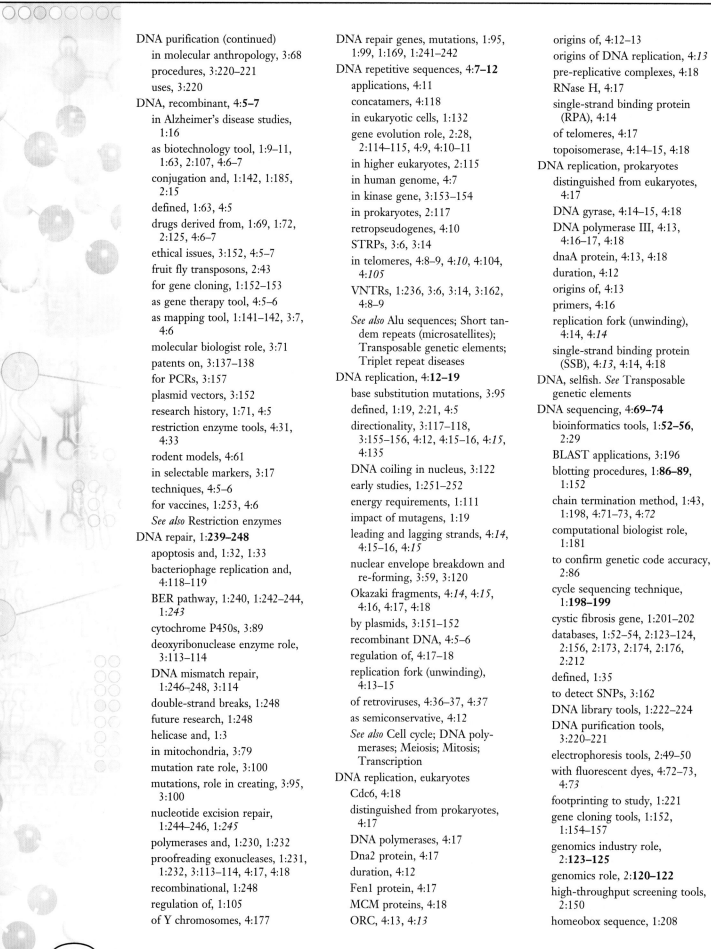

Necrosis, defined, 1:31, 4:169

Negative selection marker systems, 3:16

Neisseria gonorrhea, antibiotic resistance, 1:26, 1:27, 1:28

Neisseria meningitidis, genome characteristics, 1:142

Nematodes
  as biopesticides, 1:57
  defined, 1:57, 4:62
  holocentric chromosomes, 1:114
  plant resistance to, 1:149
  See also Roundworms

Neonatal, defined, 1:3

Neoplasms
  defined, 1:92
  See also Tumors

Neo^r gene, targeting studies, 4:125–126

NER (nucleotide excision repair), 1:244–246, 1:245, 1:247

Neural tube defects (NTDs)
  as multifactorial disorder, 1:77–78
  prenatal genetic testing, 3:184, 3:185

NeuroD1/beta2 genes, and diabetes, 1:211

Neurofibromatosis
  Alu sequences and, 4:146
  genomic screening, 2:168
  pleiotropic effects, 3:153, 3:154
  tumor suppressor genes, 4:153

Neuroimaging, defined, 1:40

Neurological crises, metabolic diseases, 3:42

Neurological, defined, 3:42

Neurological disorders
  mitochondrial diseases, 3:54
  Parkinson's disease, 1:228, 2:5–6, 3:190, 4:159, 4:160–161
  polyglutamine disorders, 4:151

Neurons
  defined, 1:15, 3:45
  FMR-1 protein and, 1:76
  membrane transport, 1:109

Neurospora, biochemical genetics studies, 3:103

Neurotransmitters
  defined, 1:5, 4:85
  for dopamine and serotonin, 1:5, 1:39–41, 1:41
  signal transduction role, 4:85

New Synthesis, 3:32

Newborn screening, 2:98–99
  congenital adrenal hyperplasia, 3:176, 3:177
  cystic fibrosis, 3:177

galactosemia, 3:44
  hemoglobinopathies, 3:177
  hypothyroidism, 3:176–177
  medium-chain acyl-CoA dehydrogenase deficiency, 3:176
  phenylketonuria, 3:42–43, 3:176, 3:218–219

Newborn screening tests
  DNA- *vs.* non-DNA based, 3:176–177
  Guthrie, 3:176
  hemoglobin electrophoresis, 3:176

Neyman, Jerzy, 4:97

NF1 and *NF2* genes, 3:130, 3:153, 4:153

Nic sites, on DNA, 3:151, 3:151

Nickel, hyperaccumulators of, 1:61

Nicotine addiction. *See* Smoking

NIH (National Institutes of Health)
  genetic discrimination studies, 2:94
  Human Genome Project role, 2:173–174, 2:176

Nirenberg, Marshall, 1:193, 1:252–253

Nitrite preservatives, as mutagens, 3:88

Nitrogen fixation, transgenic plants, 2:106

Nobel Prizes
  bacterial genetics, 1:183–184
  blood group systems, 1:82
  chromosomal theory of inheritance, 3:76, 3:80
  DNA sequencing methods, 2:172, 4:64–65
  DNA structure, 2:171, 4:172
  Kuru transmission, 3:189
  multiple, Fred Sanger, 4:64
  oncogene research, 4:38
  operon discovery, 3:131
  PCR invention, 2:172, 3:91, 3:154
  phage genetics, 1:204
  prion hypothesis, 3:187
  radiation-induced mutations, 3:81
  recombinant DNA, 1:71, 1:72
  restriction enzymes, 1:71
  retroviral reverse transcriptase, 4:35, 4:40
  RNA catalysts, 4:44
  site-directed mutagenesis, 3:91
  transposable genetic elements, 3:10, 3:22

Nomenclature, 3:**106–108**

Noncoding DNA sequences. *See* Introns

Nondisjunction, 1:*120*, 3:**108–112**
  aging and, 3:112
  and aneuploidy, 1:257, 3:110–111, 3:166
  and chromosomal mosaicism, 3:79
  defined, 1:121
  early studies, 1:131
  fatal *vs.* non-fatal conditions, 3:111
  mechanism, 3:108–109, 3:*109*
  spindle checkpoint errors, 3:111–112

Non-insulin-dependent diabetes mellitus (diabetes type 2), 1:209, 1:210, 1:212, 3:125, 3:154

Nonpolar, defined, 1:116, 3:200

Nonprocessed pseudogenes, 3:210–211, 3:*210*

Nonsense mutations, 2:127, 3:153

Nontemplate strands, 4:107

NOR-staining of chromosomes, 1:128

Northern blotting
  defined, 1:86
  procedure, 1:86–88

Novartis, drug development, 2:124

NPY mice, 1:5

NTDs (neural tube defects)
  as multifactorial disorder, 1:77–78
  prenatal genetic testing, 3:184, 3:185

N-terminus, of amino acids/proteins, 3:181, 3:197–198, 3:207

Nuclear hormone receptor superfamily
  anabolic steroids, 2:163
  function, 2:161–163
  mutations, consequences, 4:115
  number, in humans, 2:161
  number, in roundworms, 2:161–162
  RXR partners, 2:160

Nuclear lamina, 3:120, 3:121, 3:*121*, 3:125

Nuclear localization signals, 3:126

Nuclear magnetic resonance, 3:71, 4:48

Nuclear membranes (envelopes), 1:*105*
  breakdown and reformation, 3:59, 3:120
  defined, 1:139
  structure and function, 1:112, 1:139, 3:120, 3:*120*, 3:122, 3:*123*, 3:125

Nuclear pores, structure and function, 3:120, 3:*121*, 3:122–123, 3:*123*, 3:125–126

in situ hybridization tools, 3:*129*

transcription factors as, 3:*128*, 3:130

Oncogenes, specific

breast cancer, 1:91–92

cellular (*c-onc*), 4:38

colon cancer, 1:167–170

Ras, 1:99–100, 3:128, 3:130

v-*erbB*, 3:130

v-*fms*, 3:130

viral (*v-onc*), 4:38

v-*sis*, 3:130

Oncogenesis, defined, 1:92

Oncoproteins, viral, 4:155

Oocytes

age of, 4:25

collecting, assisted reproduction, 4:21

defined, 1:121, 2:45, 3:24, 4:21

forced maturation of, 4:25

frozen, 4:25

fruit fly development, 1:206–207, 1:*207*, 2:45

primary, 3:24, 3:29

Open reading frames (ORFs)

of cDNA, 1:154

defined, 2:53, 4:145

of DNA transposons, 4:9, 4:145

of pseudogenes, 3:209

Operons, 3:**131–135**

discovery, 3:131

function, 1:142, 2:15, 3:105

functional relationships, 3:132

gene clustering, 3:132

lac, 2:15, 3:131–135, 3:*132*, 3:*133*

operator regions, 3:132, 3:134

promoter regions, 3:132, 3:134

regulation mechanisms, 3:134–135

transcription of, 3:134–135

Ophioglossum reticulatum, polyploidy, 2:114

Opportunistic agents (pathogens), 2:151, 2:155

Opsin proteins

role in color vision, 1:171

signal transduction role, 4:89

Oracle databases, 1:55

ORC (origin recognition complex), 4:13, 4:*13*, 4:18

ORFs. *See* Open reading frames

Organelles, defined, 2:108, 3:79, 4:42

Organic acid metabolism, disorders, 3:40

Organic, defined, 3:88

Orgel, Leslie, 2:24

OriC point, prokaryote chromosomes, 1:141

Origin of replication (ori) sequences, 1:152–153, 3:151, 3:*151*, 4:12–13, 4:*13*

Origin recognition complex (ORC), 4:13, 4:*13*, 4:18

Ornithine transcarbamylase deficiency, 1:69, 3:45

Orthologs, 2:158

Oryza sativa. See Rice

Osmotic, defined, 2:11

Ossification

defined, 2:130

endochondral, 2:130

Osteoarthritis, 2:131–132

defined, 2:131

Osteogenesis imperfecta (OI), 2:132, 2:201

Osteoporosis

aging and, 1:7

defined, 1:1, 2:90

genetic counseling for, 2:90

progeria and, 1:1

Ostreococcus tauri, size, 1:108

Outcrossing

Arabidopsis thaliana, 1:35

vs. inbreeding, fitness, 2:146

of maize, 3:9

Mendel's studies, 1:147

Ovalbumin, function, 3:199

Ovarian hyperstimulation syndrome, 4:22

Ovaries

cancer, 1:93, 1:169, 2:102, 3:218, 4:22

development, 1:21, 1:*22*, 4:78–79, 4:*79*, 4:80–81

function, 2:*34*, 2:160, 3:24, 3:60

removal, and breast cancer, 1:90

tumors, and breast cancer, 1:91

Overdominance hypothesis, for heterozygote advantage, 2:147

Overlapping genes, 2:85–86, 3:**135–136**

Oviduct, defined, 1:163

Ovulation

defined, 1:163

drugs to enhance, 4:21, 4:22, 4:24, 4:28

and meiosis, 2:34–35, 3:29

pregnancy role, 4:19

Ovules

defined, 1:35

and Mendelian ratios, 1:130

Ovum

defined, 1:160

personhood of, 4:27–28, 4:29

Oxidation

defined, 1:63

errors, DNA damage, 1:*240*, 1:242

of methionine, 1:63–64

Oxidative metabolism, 3:41–42

Oxidative phosphorylation

ATP production, 3:52

by mitochondria, 3:51–52, 3:55–56, 3:57

reactive oxygen species, 3:52

Oxygen

consumed by mitochondria, 2:194

as electron acceptor, 2:13, 3:41–42

globin carriers, 2:136, 2:137–138

reactive species, as mutagens, 3:52, 3:100

Oxytocin

and maternal condition, 4:29

post-translational control, 3:181

P

P values, in statistical analyses, 3:194–196, 4:97

P22 virus, 4:118

P53 gene

and cancer, 1:167–168, 2:5, 4:154–155, 4:169

mutations, 1:95, 3:98, 3:130

Pac (puromycin-n-acetyl-transferase) gene, as selectable marker, 2:72

PAGE (polyacrylamide gel electrophoresis), 2:46–47

2-D, 2:48–49, 2:*48*, 3:207–208

SDS, 2:47–48

PAH (phenylalanine hydroxylase), 1:82, 2:55, 2:99, 3:42–43, 3:176, 4:5

PAHs (polycyclic aromatic hydrocarbons), 1:59, 1:99

Pairwise differences, in DNA sequences, 3:167–168, 3:*168*

PAIS (partial androgen insensitivity syndrome), 1:25

PAI-1 (plasminogen activating inhibitor 1), 1:8

Palindromic DNA sequences, 1:218

Palister-Hall syndrome, 2:130

Pancreas, insulin production, 1:209

Pancreatic disorders

cancer, 1:93

cystic fibrosis, 1:200, 1:201

mitochondrial diseases, 3:54

Pancreatic elastase, 2:27

Pandemics, defined, 1:28

Pangenes, 1:130

Phenocopies, defined, 2:61

Phenotypes
Bombay, 2:8, 2:9
defined, 1:21, 2:8, 3:9, 4:1
endophenotypes, 4:2, 4:2
Hardy-Weinberg predictions from, 2:135
mutation manifestations, 1:35, 1:75, 1:155, 1:156–157, 2:43
QTL effects on, 4:2–3
simple vs. complex, 1:177, 2:55, 2:60–61
See also Genotype, phenotype and; specific diseases and traits

Phenylalanine (Phe)
genetic code, 2:85, 4:137
and PKU, 3:42–43

Phenylalanine hydroxylase (PAH), 1:82, 2:55, 2:99, 3:42–43, 3:176, 4:5

Phenylketonuria (PKU)
as frameshift mutation, 4:5
gene-environment interactions, 2:55–56
Guthrie test, 3:176
maternal condition and, 1:82
newborn screening, 2:99, 2:119, 3:42–43, 3:176, 3:218–219
statistics, 1:75
symptoms and treatment, 2:202, 3:40, 3:42–43, 3:176

Phenylthio-carbamide (PTC), ability to taste, 2:191

Pheromones, defined, 2:158

Philadelphia chromosome, 4:154

Phimthop, Thawiphop, 1:78

PhiX174 bacteriophage, 3:135, 4:64

Phosphatase enzymes
in DNA sequencing, 4:70
post-translational phosphorylation by, 3:178
signal transduction role, 4:87

Phosphate groups
defined, 1:105, 4:70
and DNA structure, 1:105, 2:50, 2:51, 3:115, 3:116, 3:117

Phosphatidylcholine, function, 1:242

Phosphodiester bonds
defined, 1:244, 3:112
and DNA structure, 1:216, 1:216, 1:244, 3:117–118, 3:118
hydrolysis of, 3:113, 3:114
lyase reactions, 3:113
and RNA structure, 4:47, 4:49

Phosphoglycerate kinase 2, 2:28

Phospholipase C, signal transduction role, 4:89

Phospholipids, in cell membranes, 1:109, 2:12, 3:55

Phosphorimidazolide, and ribose-phosphate backbone, 2:23, 2:24

Phosphorous, toxicity, 1:61

Phosphorylation
of antibiotics, 3:17
by ATP, 3:178, 3:204, 4:87
by CREB, 4:89
defined, 1:53, 2:62, 3:51, 4:86
DNA replication initiation role, 4:18
by GDP and GTP, 4:88, 4:89–90
of proteins, 1:105–107, 2:62, 3:20, 3:178, 3:180, 3:203–204, 3:205, 4:86–87, 4:87, 4:111
of resistance genes, 3:16–17
of RNA polymerase, 4:113
RTK pathway, 4:87–88
signal transduction role, 4:86–87, 4:87
See also Kinase enzymes

Phosporimadazote, nucleotide linking role, 2:24

Photopigments, role in color vision, 1:170–172

Photoproducts, as mutagens, 1:244

Photoreceptors
Arabidopsis research, 1:36
role in vision, 1:170–172

Photosynthesis
by chloroplast ancestor, 2:195
chloroplast role, 1:112, 2:108, 2:194

Phototrophs, described, 2:13

Phylogenetic Species Concept, 1:189

Phylogenetic/phylogeneticists, defined, 1:36, 4:41, 4:146

Physical maps (genes), 1:155, 1:156

Physician, defined, 3:147

Physician scientists, 3:147–148

Phytochelatins, 1:60

Phytophthora infestans, potatoes resistant to, 2:108

Phytoremediation, 1:60–61

Picornaviruses, structure, 4:165

Pigs
cloned, 4:143
knock-out, for organ transplants, 4:140, 4:142–143

Pilus, sex, 1:183, 1:183, 1:184, 2:15, 3:151

Pineapples, genetically engineered, 2:107

PIP-2 lipids, signal transduction role, 4:89

PIQ (performance ability) intelligence scores, 2:208

Pituitary glands
and endocrine disorders, 2:129
function, 2:160
suppressing, assisted reproduction, 4:21

PKA regulatory II$_\beta$ mice, 1:5

PKA (protein kinase), signal transduction role, 4:89

PKC (protein kinase C), signal transduction role, 4:89

PKC$_\epsilon$ mice, 1:5

PKU. See Phenylketonuria

Placenta
chromosomal mosaicism, 3:79
stem cells, 4:30

Plague, as bioterrorism tool, 1:69–70

Plant genetic engineers, 3:**149–150**

Plant molecular farming, 4:134–135

Plaque, and cardiovascular disease, 1:102

Plasma membranes
defined, 1:113, 3:202, 4:89
of eubacteria, 2:12
structure and function, 1:105, 1:109, 1:113, 3:202

Plasmids, 3:**150–153**
cleaved by restriction enzymes, 4:33
Col, 3:151
conjugative (fertility factors), 1:140–142, 1:182–185, 1:184, 2:117, 3:151
defined, 1:27, 1:222, 3:16, 4:128
degradative and catabolic, 3:151
for DNA vaccines, 1:254–255
electrophoresis to separate, 2:49
as gene therapy vectors, 2:78
inheritance of, 2:198, 3:151
mobilization, 3:152
promiscuous, 1:183, 1:185
in recombinant organisms, 1:71, 2:15, 3:152
relaxed, stringent, and incompatible, 3:151
replication of, 1:152–153, 2:15, 3:151–152
resistance (R), 1:27, 1:140, 1:185, 2:117, 2:198, 3:16–17, 3:151, 4:69
structure and function, 1:140–142, 1:144, 2:14–15, 2:116, 3:150–151
Ti, 2:107
of transgenic bioremediators, 1:61
as vectors, 1:222, 2:107, 3:91, 3:152
virulence, 3:151

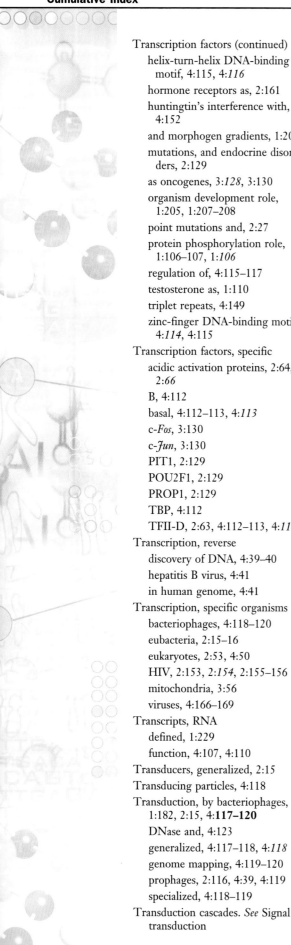